CANADA'S RANGERS
Selected Stories 1942-2012

CANADA'S RANGERS

Selected Stories 1942-2012

Compiled & Introduced by
P. Whitney Lackenbauer

CANADIAN DEFENCE ACADEMY PRESS

Canadian Defence Academy Press
PO Box 17000 Stn Forces
Kingston, Ontario K7K 7B4

Produced for the Canadian Defence Academy Press
by 17 Wing Winnipeg Publishing Office.
WPO 30840

Library and Archives Canada Cataloguing in Publication

Canada's Rangers : selected stories, 1942-2012 / compiled and introduced by P. Whitney Lackenbauer.

Produced for the Canadian Defence Academy Press by 17 Wing Winnipeg Publishing Office.
Includes index.
Available also on the Internet.
Issued by: Canadian Defence Academy.
ISBN 978-1-100-21570-9
Cat. no.: D2-312/2013E

1. Canada. Canadian Armed Forces. Canadian Rangers--History. 2. Canada--Armed Forces--Native peoples. 3. Canada--Armed Forces--Canada, Northern. 4. Canada--Armed Forces--Operations other than war. 5. Canada--Armed Forces--Guard duty. 6. National security--Canada. 7. Search and rescue operations--Canada. 8. Disaster relief--Canada. I. Lackenbauer, P. Whitney II. Canadian Defence Academy III. Canada. Canadian Armed Forces. Wing, 17

UA602 C36 C36 2013 355.3'70971 C2013-980019-0

Printed in Canada.

1 3 5 7 9 10 8 6 4 2

– CONTENTS –

CONTENTS

CONTENTS

CONTENTS

CONTENTS

CONTENTS

– PREFACE –

I am delighted to introduce this collection of stories about the Canadian Rangers – an important part of the defence team who have served for sixty-five years as the "eyes and ears" of the Canadian Forces (CF) in sparsely settled northern, coastal and isolated parts of Canada. As a sub-component of the CF Reserve, the Canadian Rangers provide patrols and detachments for employment on national-security and public-safety missions in areas that cannot be conveniently or economically covered by other elements or components of the CF. These dedicated, knowledgeable members of the Army reflect the diversity of their home communities. The Canadian Rangers' red sweatshirts have become a visible image of sovereignty, security, and stewardship in remote regions, and high rates of Aboriginal participation in the Canadian Rangers play an important role in advancing public recognition of Canada's Inuit, First Nations and Métis as vital members of the CF.

This volume provides readers with an overview of the history of the Canadian Rangers and also introduces them to the wide array of contributions that the Rangers have made, and continue to make, to Canada and to their home communities. Historian Whitney Lackenbauer, the leading civilian expert on the Canadian Rangers, has compiled a wide range of newspaper and magazine articles that trace their origins in the Pacific Coast Militia Rangers from 1942-45, through their formal establishment as a national organization in 1947 to present. The journalists and commentators whose articles are reprinted in this book reveal how the Canadian Rangers protect Canada's sovereignty by reporting unusual activities or sightings, collecting local data of significance to the CF, and conducting surveillance or sovereignty patrols. They also make important local contributions, participating in Search and Rescue (SAR) operations, providing support in response to natural or man-made disasters and humanitarian operations, and assisting federal, provincial, territorial, and municipal authorities in times of need. Their value as an operational resource for the Canadian Forces is clear in the following stories, as is their value to their communities.

Most importantly, these articles reveal experiential knowledge, insights and testimonies from the Canadian Rangers themselves, as well as members of their communities and other CF personnel who have served with them. Although these stories do not represent the official views of the Department of National Defence or of the Canadian Forces, they express a richness of opinion and information that can assist members of the CF who are preparing to work with the Rangers. Equally important, these stories are an important affirmation for the Canadian Rangers that their ongoing contributions are appreciated and recognized outside of the military.

In closing, I wish to reiterate the importance of this book for the Canadian Rangers and those Canadians who serve in, and interact with, the profession of arms in Canada. It is a fitting tribute to the collaborative relationships and achievements of the Canadian Rangers, their home communities, and other elements or components of the CF who contribute to safety, security and defence in northern, coastal and isolated areas of our country.

Major-General P.J. Forgues
Commander
Canadian Defence Academy

– FOREWORD –

These Are Canada's 'Very Special Forces' of the North

Back in 1994 I was a staff reporter with the *Globe and Mail* and I had just flown into Fort Albany, a small and remote Cree community on the cold, wind-swept coast of James Bay in Northern Ontario.

I arrived at a small house that was used by transient visitors to claim my bunk bed, only to find a Canadian army rucksack and other military gear on it. Beds being in short supply I went looking for the rucksack's owner. And that's how I became a Canadian Ranger.

The owner of the kit was Captain David Scandrett who was in the process of establishing the Rangers in Northern Ontario. He relinquished the bed and we became friends. When I retired from *The Globe* he convinced me I should become a Ranger and use my 38 years of experience as a journalist to write about the Rangers. I have done that happily for the last 12 years, writing, as the public affairs ranger for the 3rd Canadian Ranger Patrol Group, for both military and civilian publications.

I am proud to see several of my stories in this remarkable collection of articles about the Rangers, which begins with stories about the founding of the Pacific Coast Militia Rangers in 1942.

The Canadian Rangers of today are a uniquely Canadian military organization. No other country has anything quite like them. As part-time army reservists they are true citizen soldiers in small communities all across Canada's remote and isolated regions. They operate in some of the harshest weather and challenging terrain of anywhere in the world. Their selfless service has won them medals, awards and commendations, and praise from some of the most experienced soldiers in the world.

The articles tell about their contributions in the Second World War when there were concerns the Japanese might invade British Columbia. They give gripping accounts of search and rescue missions; evacuations of remote communities threatened by flood, forest fire, and tainted water; and how Rangers reacted swiftly to save lives when an avalanche devastated a community in Quebec.

There are stories about them as skilled mentors to other members of the Canadian Forces when they venture into the forbidding North, how they guide and teach other soldiers how to survive and move safely on the land, ice, and water.

High in the Arctic they play a vital role in preserving Canada's sovereignty. They inspect radar sites and participate in both routine and epic patrols, such as when they led other soldiers to the Magnetic North Pole in 2002.

These stories tell of the Rangers' pride in being masters in their own land, serving Canada and helping their communities. They act as role models for their youth through the Junior Canadian Ranger program. And, as these stories show, they seem to always display an almost constant good humour, even in the most rigorous and challenging of circumstances and conditions.

In 2002, a small group of Canadian Rangers from Northern Ontario went to the U.S. Army base at Fort Knox in Kentucky to participate in a major military exercise. A reporter for the *National Post* asked me at the time about the difference between the Canadian Rangers and the U.S. Army Rangers, an elite force among the world's best-trained fighting units.

"The American Rangers are Special Forces. Our Rangers are *very* special forces," I told him. I asked him not to attribute the quote to me. I wasn't sure how the Canadian Forces would appreciate me saying it. So the reporter said it was "a quip by a Canadian Forces officer."

Today I'm proud to admit I said it. The Canadian Rangers are, indeed, very special forces, because, as the stories in this book show, that is what they truly are. As a former commanding officer of mine once told a reporter: "The Rangers are a unique unit and you should be proud of these guys. There's just not enough Rangers to go around."

A word of advice: because these are mainly articles in newspapers, magazines, and newsletters many of them have repeated references to the founding of the Rangers and explanations about the Rangers' basic role in the Canadian Army. Don't worry about it. Skip those paragraphs if you want to and carry on reading.

Your reward will be some wonderful stories about a dedicated group of Canadian men and women who are proud to serve the Canadian Army and their country with distinction.

Sergeant Peter Moon, MMM, CD
CFB Borden, Ontario
2012

– Acknowledgements –

I am particularly grateful to the following authors and publishers for their permission to reprint newspaper and magazine articles:

- Canadian Defence and Foreign Affairs Institute / Sarah Magee
- Canadian Press / Lora DiPoce
- Globe and Mail / Marilyn Hertz
- Klondike Sun / Dan Davidson
- Legion Magazine
- Nickel Belt News / John Barker
- Northern News Services / Derek Neary
- NTI / Kerry McCluskey
- Nuntsiaq News / Jim Bell
- Peter Moon / 3 CRPG
- Postmedia Network / Cheryl Minnis
- The Labradorian / Barbara Dean-Simmons
- The Northern Pen / Juris Graney
- The Terrace Standard / Rod Link
- The Western Star / Troy Turner
- Timmins Times / Len Gillis
- Torstar Syndication Services / Joanne MacDonald
- Up Here Publishing / Katharine Sandiford
- Vancouver Sun / Adrian Mitescu
- Victoria Times-Colonist
- Whitehorse Star / Jackie Pierce

I hope that this collection will help to promote the Canadian Rangers and make these stories about their contributions more accessible to Rangers, their communities, and Canadians in general.

A special thanks to Lisa Beiler, whose research assistance was instrumental to the completion of this book. Similarly, Sergeant Peter Moon, MMM, CD, 3 CRPG's public affairs Ranger, offered valuable suggestions on how to improve the collection and contributed a foreword. I am grateful to the Canadian Rangers National Authority for supporting this project in spirit from the onset, to ArcticNet for funding through "The Emerging Arctic Security Environment" project and to the Walter and Duncan Gordon Foundation – Munk School for Global Affairs Arctic Security Project for facilitating ongoing research on the Canadian Rangers, and to Peter Kikkert and Jennifer Arthur-Lackenbauer for valuable editorial comments.

The text is accompanied by photographs taken by Canadian Forces photographers (and some that I have taken over the last decade). Thanks to various Canadian Ranger Patrol Groups for providing selections from their photograph collections and to the Department of National Defence for permission to reproduce images from its extensive photographical holdings.

– Introduction –

The Canadian Rangers are popularly recognized as Canada's "eyes and ears" in remote regions. Since 1947, their official mission has been "to provide a military presence in sparsely settled northern, coastal and isolated areas of Canada that cannot conveniently or economically be provided for by other components of the Canadian Forces." Over the last six decades, the tasks that they perform in support of this mission have become more complex. Their initial focus was national *security* – protecting their communities from enemy attack in the early Cold War. By the 1970s, their responsibilities became directly linked to the armed forces' role in support of Canada's *sovereignty* in the Arctic. Since the 1990s, the Rangers have also played a more visible nation-building and *stewardship* role in remote regions across Canada. They represent an important success story for the Canadian Forces as a flexible, inexpensive, and culturally inclusive means of having "boots on the ground" exercising Canadian sovereignty and conducting or supporting domestic operations. As a bridge between cultures and between the civilian and military realms, the Rangers represent a successful integration of national security and sovereignty agendas with community-based activities and local stewardship. This practical partnership, rooted in traditional knowledge and skills, promotes cooperation, communal and individual empowerment, and cross-cultural understanding.

Through my research into the history of the Canadian Rangers over the last decade, I have gathered a large collection of newspaper and magazine articles. These documents provide a sampling of the many activities in which the Rangers have been involved, reveal how the organization has evolved, and affirm the positive influence that the Rangers have throughout Canada. I hope that this collection serves as a celebration of the Canadian Rangers' contributions to their communities and their country over the last seven decades.

Historical Overview

The first group of articles relates to the Pacific Coast Militia Rangers (PCMR), a sub-component of the Army Reserves (part-time citizen soldiers) created during the Second World War. During the war, Canadians realized that they no longer lived in a "fireproof house" (as Senator Raoul Dandurand described Canada in the 1920s). In a total war, isolation no longer suggested security; instead, it represented vulnerability. Facing the Japanese threat in the Pacific, terrified British Columbians pushed the federal government to improve its defences along the west coast. This led to the formation of the PCMR in 1942. Based on the model of the British home guard, the Ranger organization allowed BC men who were too old or too young for military service, or who were engaged in essential industries, to contribute to home defence. They were not issued vehicles or regimental equipment - Rangers were expected to use their own. Apart from a sporting rifle, some ammunition, an armband, and eventually a "Dry-bak" uniform suited to the coastal climate, the Rangers were self-sufficient. Their defence duties were designed to "not conflict with their normal civil employment" unless a state of emergency arose and they were called out on active service. Basically, they were expected to know their local areas, to act as "eyes and ears" for the military reporting any suspicious vessels or activities, and to do what they could to repel an enemy invasion in concert with other military forces. "The Ranger movement" proved popular. By 1943, there were 15,000 Rangers from all walks of life, from fish packers to cowboys. When the war ended, however, so did the PCMR. The organization was stood down in the fall of 1945.

As the wartime alliance between the democratic West and the communist East unravelled at the end of the Second World War, Canada became a potential battlefield in a future superpower conflict. In 1947, the USSR developed an intercontinental bomber. Two years later, the Soviets exploded their first nuclear bomb and the threat of a continental attack became more serious than ever before.

The Canadian military recognized that it could not feasibly station large numbers of regular soldiers in northern and remote regions of the country. Nevertheless, the military needed to have people serving as its "eyes and ears" beyond the main population belt, so officials resurrected the Ranger concept in 1947. The Rangers represented a cost-effective solution to Cold War sovereignty and security concerns that drew upon existing human resources in local areas. The plan was to recruit individuals who would not appeal to other military units because of age, health or employment reasons, and would therefore remain in their home communities in both war and peace. With little training and equipment, the Rangers could act as guides and scouts, report suspicious activities and – if the unthinkable came to pass – delay enemy advances using guerrilla tactics. The only equipment issued to Rangers at that time was an obsolescent .303 Lee Enfield, 100-200 rounds of ammunition annually, and an armband. From the onset, the force structure was decentralized because variations in geography and roles made it impossible to create a "standard establishment." Each Ranger platoon was operated and administered locally.

The next set of articles describes the formation of the Rangers, as well as their involvement in army exercises during the 1950s. Annual re-supply and training visits by Regular Force Ranger Liaison Officers (RLOs) provided opportunities for cross-cultural contact through that decade (as Captain Ambrose Shea's fascinating articles reveal), but by the 1960s the Rangers were peripheral to Ottawa's defence plans. Northern residents with armbands and rifles were no match for Soviet bombers carrying nuclear weapons. Defence officials turned to technological marvels like the Distant Early Warning (DEW) Line to protect the continent, but even these would be ineffective in the face of intercontinental and submarine-launched ballistic missiles. The Rangers cost next to nothing, however, which meant that they survived. This "Shadow Army of the North" received little direction, sporadic re-supply, and no training. Apart from Newfoundland and Labrador and a few northern communities, the Rangers were inactive by 1970, as Major Bill Stirling described when he visited platoons across the country.

The federal government expressed renewed interest in the Canadian North in the wake of the American icebreaker *Manhattan*'s voyages in 1969-70. The new Arctic "crisis" had little to do with an increased Soviet threat, and everything to do with Canadian sovereignty in the North. Prime Minister Pierre Trudeau turned to the Canadian Forces to take symbolic steps to assert control, promising increased arctic surveillance and placing more emphasis on Arctic training for southern troops.

Who would train them? Only people who lived in remote regions had the requisite expertise to act as guides and survival instructors. As a result, Northern Region Headquarters made modest progress in resurrecting northern Ranger units. The simple fact that the Rangers already existed as an "officially constituted" element of the CF and asserted sovereignty at a minimum cost were important considerations at a time when the government was unwilling to commit a lot of personnel or money to military matters. Although the bold proposals to reorganize the Canadian Rangers as a Primary Reserve unit were never implemented for financial reasons, the army did provide training to groups of Inuit and Dene Rangers in the 1970s. These activities proved highly popular in communities, offered the military permanent contacts across expansive northern and coastal regions, and built relationships so that the Rangers could serve as guides and advisors for military units exercising in remote regions. By the early 1980s, the Rangers were again active in the territorial north, northern Quebec, and along the eastern seaboard in Newfoundland and Labrador.

When the US Coast Guard icebreaker *Polar Sea* pushed through the Northwest Passage in 1985, the perceived challenge to Canadian sovereignty renewed demands for a bolder Canadian presence in the Arctic. The Mulroney government made a host of promises to deal with sovereignty, from acquiring nuclear submarines to building a Polar Class icebreaker. Concurrently, the Canadian Rangers drew increasing attention as an important source of sovereignty and security. Up to the mid-1980s, the Department of National Defence's official assessments of the Canadian Rangers had focused on military utility, but in a changing political climate other aspects of the organization heightened its attractiveness. Military activities in the North could no longer be divorced from domestic socio-economic, cultural, and environmental health issues. Aboriginal leaders repeatedly called for the demilitarization of the arctic for social and environmental reasons, construing a military presence as a threat to their peoples' security. On the other hand, they (and the media) began to emphasize the socio-political benefits of the Rangers in Aboriginal communities.

Most of the federal government's promised investments in Arctic defence evaporated with the end of the Cold War, lost in the midst of more pressing national priorities. The promise to expand the Canadian Rangers, however, was kept. The organization underwent dramatic growth. This was all the more remarkable given that the 1990s was a decade of fiscal and personnel downsizing in the Canadian Forces more generally. Nevertheless, the Rangers, and particularly the high percentage of Aboriginal people in the organization, were a success story. As the articles in this book reveal, most political and national media attention fixated on Inuit participation in the far north, where the Canadian Rangers retained their appeal as an inexpensive, culturally-inclusive, and visible means of demonstrating Canadian sovereignty. As a result, Ranger budgets increased and the military expanded their presence and enhanced their capabilities "north of 60." The organization also enjoyed more growth across Canada, consolidating the Ranger footprint along the Atlantic and Pacific Coasts, in Quebec, and returned to northern Ontario. Most new growth was directed towards Aboriginal communities, reflecting the importance of building and maintaining Aboriginal-military partnerships. Furthermore, journalists applauded the Rangers' role not only in teaching the military but also in encouraging elders to share their traditional knowledge *within* Aboriginal communities. The creation of the Junior Canadian Rangers in 1998 was a novel initiative that enhanced the Rangers' social and cultural value in remote regions.

At its core, the Canadian Ranger organization works because it is built upon trust. Over time, Ranger Instructors have forged strong relationships with Rangers across the country. Various articles in this book reveal how and why the flexible, culturally-appropriate approaches to Ranger training accommodate the diverse ways and needs of communities across Canada. It also helps to explain why, by the end of the twentieth century, the Rangers had come of age. They emerged from the shadows to play an increasingly prominent role in demonstrating Canadian sovereignty, contributing to Canadian security and safety, and serving as an important example of stewardship in remote regions.

The Twenty-First Century

The Rangers have gone from a relatively unknown military formation to "sovereignty soldiers" undertaking highly publicized enhanced sovereignty patrols across the high arctic, supporting military activities in all provinces and territories except for Nova Scotia, Prince Edward Island and New Brunswick, and performing a wide range of local activities. As the articles in this book demonstrate, the Canadian Ranger mystique gained new lustre in the national spotlight. Their durability and survival skills assumed almost mythic proportions in media stories. The Rangers exemplified the military's positive cooperation with Canadians living in remote areas, and in particular with Aboriginal communities. Furthermore, the organization proved that the military could successfully integrate national security and sovereignty agendas with community-based activities and local management. It was a practical partnership rather than shallow "consultation," rooted in community-based monitoring using traditional knowledge and skills.

The Canadian Rangers were organized into five Canadian Ranger Patrol Groups in 1998, and their coverage of Canada's remote northern and coastal regions is extensive. The figures, as of 14 November 2012, are as follows:

- 1 CRPG is located in northern Canada with a total of 59 patrols and 1617 Rangers in the Yukon, the Northwest Territories, Nunavut, and one in northern British Columbia;
- 2 CRPG is located in Quebec with 25 patrols and 731 Rangers;
- 3 CRPG is located in northern Ontario with 20 patrols and 610 Rangers;
- 4 CRPG is located in British Columbia and the northern Prairie provinces, with 42 patrols and 1054 Rangers; and
- 5 CRPG is located in Newfoundland and Labrador, with 32 patrols and 896 Rangers.
- National totals are 178 Canadian Ranger patrols and 4908 Canadian Rangers.

The Rangers' expertise in operating in their homelands contributes to the defence of North America *and* to the protection of their communities. The Rangers' updated role, mission, and tasks take into account the broad notion that "protecting Canadians" includes helping relatives and community members in emergencies or disasters (whether human-made or natural) as well as their traditional role as the CF's "eyes and ears."

Various articles in this book reveal how growing concerns about climate change, the opening of the Northwest Passage, global demands for Arctic resources and security in the post-911 world have conspired to put the Arctic back on the national and international agenda. Popular perceptions that other countries challenge Canadian sovereignty have drawn renewed attention to sovereignty operations and the importance of the Rangers as Canada's "boots on the ground." Enhanced Sovereignty Patrols, now run annually as *Nunalivut* operations, exercise sovereignty in parts of Canada's uninhabited Arctic archipelago seldom visited by human beings. The long-range patrols have become increasingly ambitious over the last decade, revealing how the Rangers have gone from providing a quiet presence in the North to offering a bold, symbolic one. As various articles in this volume attest, these activities allow the Rangers to operate in unfamiliar environments, share skills, develop relationships with other members from across the North and serve as confidence-building measures for participants. The epic nature of enhanced sovereignty patrols has pushed the Rangers' responsibilities in the Arctic far beyond their original concept of simply knowing their local areas.

Although it is grand gestures such as sovereignty patrols and joint military operations (such as Operation *Nanook* held each August) that attract high levels of media attention, the articles in this book reveal how the Canadian Rangers assist CF activities in other ways. They provide local expertise and guidance; advise and instruct other CF personnel on survival techniques, particularly during sovereignty operations (SOVOPs); support the Junior Canadian Rangers program; and provide local assistance to search and rescue and disaster relief activities. Dramatic responses to avalanches, forest fires, flooding, and water crises reaffirm the importance of the Rangers to domestic operations across the country. Furthermore, the Rangers in all regions provide a valuable service in teaching southern-based units practical, traditional survival skills during winter warfare training and other activities. These interactions encourage cross-cultural awareness and understanding. The very positive assessments of the Canadian Rangers by Regular Force and Reserve soldiers (both Canadian and international), cited in many newspaper and magazine stories, testify to the bonds formed and the Rangers' important contributions to defence.

The final Ranger task is the most general and basic – to maintain a CF presence in local communities. This is fundamental to Canada's commitment to having a military "footprint" across the country. The Rangers represent more than ninety percent of CF representation north of the 55th parallel and enjoy a special relationship with their host populations. The articles in this book demonstrate that they are far more than the military's "eyes and ears." They are an organized group to which communities turn for leadership and support.

The Canadian Rangers are a success story from coast to coast to coast, promoting sovereignty, security, safety and stewardship from the "inside out." The federal government has promised expansion, enhancement, and an investment of a quarter of a billion dollars in funding over the next two decades – a huge amount given the shoestring budget allocated to the Rangers in the second half of the twentieth century. The positive relationships that have been forged between the Rangers and other Canadian Forces elements over the last half century have produced high levels of trust, cohesion and morale. For good reason, they are a key component to Canada's sovereignty, security, and stewardship responsibilities in remote regions. The stories in this book offer a sample of their contributions to their communities, the Canadian Forces, and their country – past and present. I hope that you enjoy them as much as I do.

P. Whitney Lackenbauer
Otterville, Ontario
March 2012

Note on language:

The language in the various articles reflects the time in which they were written. Some readers may find particular terms offensive today, but I have chosen not to change them because they must be understood in the context of the time period in which they were produced. Instead, readers are encouraged to reflect on how our language and descriptions of individuals and groups have changed over the last seven decades.

– Canadian Ranger Motto –

Vigilans (The Watchers)

– Canadian Ranger Mission –

The mission of the Canadian Rangers is "to provide lightly equipped, self sufficient, mobile forces in support of the CF's sovereignty and domestic operation tasks in Canada".

– Tasks - Canadian Rangers –

The National Canadian Ranger Task List is as follows:

- Conduct and Provide Support to Sovereignty Operations:
  - Conduct, participate in and provide support to sovereignty and surveillance patrols and training in Canada
  - Conduct North Warning Site patrols
  - Report suspicious and unusual activities
  - Collect local data of military significance.
- Conduct and Provide Assistance to CF Domestic Operations in Canada:
  - Conduct coastal and inland water surveillance
  - Provide local knowledge and CR expertise (guides and advice)
  - Participate in Search and Rescue (SAR) operations
  - Provide support in response to natural or man-made disasters and humanitarian operations and;
  - Provide assistance to Federal, Provincial/Territorial or Municipal authorities.
- Maintain CF presence in the Local Community:
  - Instruct, mentor and supervise the Junior Canadian Rangers; their work with the Junior Canadian Rangers (JCR) Program, while not an "operational" task, does assist in achieving national goals through nation building and significantly improving the quality of life of young people in the most isolated areas of Canada; and
  - Participate in/support events in the Local community (such as Yukon Quest, Canada Day, and Remembrance Day, etc.)

-1-

Rangers Will Be Roughest, Toughest Men in B.C.

Vancouver Sun, 17 April 1942

District commanders of the Pacific Coast Militia Rangers, Canada's new home defense force, concluded conferences with Col. T.A.H. Taylor, OBE, MC, on Thursday and training of units throughout the province starts today.

Organization of the force was outlined to The Sun today by Col. Taylor who was named officer commanding the Ranger less than ten days ago.

In this time more than 5000 men have been enrolled as members, 35 companies have been authorized and commanders appointed, 20 more companies have started organization.

"When this force is at full strength we will have 140 units of the roughest and toughest men in B.C. available as a reception committee for any invader," Col. Taylor said to a Sun reporter.

The whole province will be divided into areas in which Ranger bands will be organized. Organization of many districts [has] been completed and regional commanders are already in the field after holding a week-long conference with Col. Taylor.

Among the district commanders are men whose names are household words among soldiers. Col. Cy Peck, VC, DSO, who commanded the famous 16th Scottish, is commander in the Sidney area. Brigadier E.J. Ross, DSO, MC, a veteran Indian army officer who pioneered in the Peace River country, will lead another unit.

Lt.-Col. Leslie Coote, former officer commanding the Westminster Regiment who has just returned from Home Guard duty in England, will be the leader in the Fraser Valley.

"We want the dominant personality in each area to be the commander," Col. Taylor explained. "We don't care if his signature is a thumb-print as long as his men will follow him.

"This is the toughest country in the world. It is not a bit of use taking a man from outside and giving him the job of commander no matter how good his record as a soldier may be.

He'd simply get lost in the bush."

This spirit of organization to fit the situation will permeate the whole Ranger movement. No hard and fast rules have been laid down to the commanders. They have been sent into the field and told to do whatever is best for the defense of their particular areas.

Col. Taylor illustrated the work which has already been done by using the Duncan-Cowichan-Nitinat area as an example. There 14 companies have been formed; nine are composed of loggers, four of farmers, and two of other residents.

"These men know the bush and every stream and hill in their communities," he said. "Better than any man sitting before a map they can tell how best that area can be defended.

"They will train where and when and however seems best to the unit leaders. Their privately owned arms will be supplemented by military weapons as circumstances demand and supplies are available."

Fund of Information

When province-wide organization has been completed, the headquarters of Pacific command will be in touch with almost every community and through the Rangers will command a fund of information which would not otherwise be available.

"If Maj.-Gen. [R.O.] Alexander wishes to know the conditions of transport which exist at any spot he should be able to get that information from the Rangers' unit within an hour. Short of weeks of work such information would otherwise be completely unavailable," said Col. Taylor.

At the Rangers' headquarters in Victoria there is a small staff in addition to field supervisors who have been chosen according to their military experience, knowledge of the bush, and personality.

The supervisors have to be men who have roughed it and toughed it. "They must have common sense and be able to talk to trappers and loggers, as well as mayors and reeves in their own terms," he said.

No Uniforms

Each member of the Rangers will have a regimental number and be signed on as a member of the militia. No uniforms will be worn but the soldiers will wear the armbands of their unit.

Col. Taylor has had as much bush and back-country experience as many of the men he will command. He has been a surveyor, timber cruiser and shingle-mill laborer, as well as a soldier. In addition he had three years' experience as a construction worker along the African Gold Coast.

– 2 –

Coast Rangers in a Class by Themselves: Prowess in the Woods Their Chief Weapon

Hal Straight

***Vancouver Sun*, 21 April 1942**

Victoria -- You can't spit tobacco juice in front of the shiny toes of a colonel and get away with it in the army--unless you are a Pacific Coast Militia Ranger. You can't report for duty covered with coal dust, or buckskin clad, or in agricultural costume, or in the imported tweeds characteristic of retired British officers--unless you are a Pacific Coast Ranger.

Brass buttons, stripes, heel-clicking mean nothing to these Rangers who make up the recently formed guerilla militia. Their only salute is to mother nature, whose complexities have won their respect years ago and by the same token are the reasons why the Pacific Rangers are so important to the defense of British Columbia.

Varying Topography

I am making a tour of several of the units of the Pacific Coast Rangers, with Lieut.-Col. Tommy (T.A.H.) Taylor. I say "Tommy" because that's what the hardy outdoor artisans call the head of the Pacific Coast Rangers.

He said before we started: "The Pacific Coast is so varying in topography, is divided into so many different zones, all of which have different conditions of growth and formation that it would take years for an army man to cope with the problems that would arise.

"We have on Vancouver Island some fine retired officers, but despite their records, if you took them 100 yards into the bush, they would get lost. So I have selected, in each zone, men who know the country, men who know each path, know every landmark, men who are men, and I don't care whether they can spell salute or can even sign their own name, so long as they know their country and if an emergency arises, such as an invasion, they can guide the military units who will naturally move in."

Perfect Choice

Lieut.-Colonel Tommy Taylor was wisely chosen for his post. He has covered a large part of Vancouver Island as a trapper, also as a timber cruiser. He has had to build his own canoe to traverse some of the rough part of the northern part of the island. He has been on construction jobs in the rugged depths of the mainland. He is an outdoor man who speaks the language of outdoor people.

When Tommy speaks to a logger, a miner, yes, a retired officer, whom he calls his "Lord Chumleys," they listen. To the logger he rattles off the rough language of the backwoods. Likewise to the miner he is a perfect choice.

But his job isn't an easy one. The Pacific Rangers were formed long before the announcement about them was made to the public. Taylor was surfeited with applications to this guerilla militia from old army friends, old fish and game friends and many other loyal British subjects. He had to be hard.

Zone System

"There was no sense me accepting some dear old fellow for a unit that needs a box of Copenhagen snuff instead of a pot of five o'clock tea. I couldn't take a fellow into the militia just because he had a nice rifle and wanted to do some shooting. I had to select men who could be men in their own territory," he explained.

Later I will describe some of these men whom he selected. But first may I disclose that the lieutenant-colonel has divided Vancouver Island and will divide the mainland of British Columbia into zones which are the same as the game districts.

"It is a natural way to zone the country. The boundaries of these zones follow the valleys between the rugged forests and mountains. And in these game zones are men who are familiar with the particular kind of country," said Taylor.

Know Shortcuts

As we toured through the neatly cultivated maize farmlands or stood in the deep forests, or climbed to the rugged peaks here on Vancouver Island this writer appreciated why "Colonel Tommy" was so emphatic about choosing men who belong to their particular part of the country.

I also could see why the Pacific Rangers will be so important in the case of invasion. They will become the generals, leading the generals of the army by the nose. They will know the shortcuts, the vital roadways to block, the waterways to divert or unleash.

The Pacific Rangers with their .30-30's, their bowie knives, their pocket compasses and their understanding of the country will justify any spitting of tobacco juice on a general's shoes."

– 3 –

The Eyes and Ears of Canada's Pacific Coast Defence: Men of the P.C.M.R. Reviving Frontier Tactics

A.G. Macdonald

***British Columbia Lumberman,* November 1942**

In civil life they pursue almost every profession, trade or occupation which the Province of British Columbia supports. They are loggers and lumbermen, farmers and fishing folk, hunters and trappers, architects, engineers and contractors, miners and ranchers, storekeepers, postmasters, clerks and accountants, coroners, and stipendiary magistrates—a cross-section of B.C. But all have one thing in common, a love for and intimate knowledge of their own particular countryside and a determination to protect it against all comers.

They are the men of the Pacific Coast Militia Rangers, and they are now, after eight months of intensive organization, almost 15,000 strong.

They are the men who know British Columbia best.

Since last March, these men have, through the efforts of their commanding officer, Lt.-Col. T.A.H. Taylor and his capable staff, been welded into an organization unique in modern Canadian military history, designed by western men to suit western conditions. We are reminded, however, by Major-General G.R. Pearkes, Pacific Command that a somewhat similar function was performed in earlier days by groups of hard-hitting irregular forces recruited from the ranks of woodsmen and Indian fighters whose stirring deeds were typified by the accomplishments of such units as Butler's and Rogers' Rangers. Basically, the P.C.M.R. organization differs from the border rangers of the past in being more highly localized for the purposes of effective home defence.

"The Pacific Coast Militia Rangers are the Rogers' Ranger mark 1942," wrote General Pearkes in a message to the P.C.M.R. in which he succinctly outlined their responsibility. "They are faced today," he continued, "by an enemy just as cunning, cruel and crafty as ever was any Shawnee or Delware Indian brave. The forests of British Columbia are just as trackless today as the Ohio wilderness was a century ago. The border needed men of high courage then, and their sacrifices were made for the benefit of future generations. Canada can still produce the same type, and there will be plenty of opportunity for the Pacific Coast Militia Rangers to show that they are worthy of the name of Ranger should any Jap invader ever dare to set his foot on British Columbia soil. The troops who are now guarding our shores come from all parts of Canada—from the cities and the prairies; they are not all familiar with the bush, and they will, therefore, need the help of those who have lived in the various types of bush in this province. It is important that mutual confidence be established between Rangers and soldiers so that each may know the qualities of the other, and each be used in his most appropriate sphere. I am proud of the Pacific Coast Militia Rangers under my command and know that they have a very definite part to play in the defence of our country. I have supreme confidence in their ability to play that part nobly when the day comes."

A Job for Specialists

A clear appreciation of the essentially different character of the Ranger organization and its peculiar function as a body of specialists familiar with this terrain in all its aspects is expressed in General Pearkes' message. In time of emergency its personnel will work in full co-operation with the regulars, but because of their special qualifications for bush campaigning, the Ranger units will enjoy a greater latitude in individual operations while doing their job as the eyes and ears of the army.

Under such headlines as "Guerilla Home Guards to be Formed in B.C" the daily press heralded the creation of the new defence body early this year, and during a brief formative period the organization was known by several designations, operating first as "Coast Defence Guards," then as "Pacific Coast Rangers," and crystallizing finally as the "Pacific Coast Militia Rangers," the word "Militia" being added by the Minister of National Defence to signify its military status.

Recognizing the special nature of the job which the P.C.M.R. was deigned to accomplish, the Department has accorded it increasingly broad powers of initiative, and beyond the inevitable routine matters of army records and administration it enjoys a free hand in organization and training which enables its officers to adapt their schedules and methods of instruction to the special requirements of the various companies, detachments and groups according to local conditions.

From the Boundary to Hudson Hope

These companies and smaller units, by the way, are now spread broadcast throughout the province. A glance at the huge map of British Columbia in the office of the O.C. shows the vast territory to be already surprisingly well blanketed with P.C.M.R. units: Colonel Taylor's map is liberally besprinkled with blue stars, each representing a P.C.M.R. group. The stars are, of course, most heavily concentrated along the coasts of the Queen Charlotte Islands, Vancouver Island and the Lower Mainland, and up the Fraser Valley, but they are also widely dispersed throughout the Southern Interior as far east as the Alberta boundary and range northwards to the shores of the Peace River, where the most remote unit is centered at Hudson Hope.

While every conceivable type of civilian enterprise is represented in the organization, loggers and lumbermen naturally predominate in the coastal area, and practically all the large logging outfits have Ranger companies, while employees of many of our big coast sawmills are to be found on the rolls of other units. As one moves away from the Coast the lumber industry is less heavily represented and farmers and ranchers become more numerous.

Elasticity of Organization

Because of its great range and the wide divergence in density of population between the Southern Mainland and the northern reaches of the province, the establishment of the P.C.M.R. is necessarily very elastic, the structure of the various units being contingent upon the number of men available in a given area and the nature of the countryside under their protection. Basically, the Rangers' organization is composed of companies, detachments and groups operating from various centres throughout B.C.

Lt.-Col. Taylor, who is on the Headquarters staff of the Pacific Command, has full responsibility for the organization and training of the P.C.M.R., and directs their activities with a staff consisting of Major W.S. Barton, assistant to the O.C., Capt. B.E.T Kennelly, training officer, and Capt. [A.F.] Watts, quartermaster.

Working out of H.Q., a staff of field supervisors, Capt. G. Baldwin, Capt. S.M. Gillespie, Capt. B.T. O'Grady, Capt. B. Harvey and Capt. J.B. Acland, travel throughout the province, assisting in the organization of new units and maintaining liaison between Headquarters and existing units.

Supplementing the work of these supervisors, eight travelling sergeant-instructors will soon be constantly in the field, bringing training facilities directly to the men in the logging camps, lighthouses, mines and villages, wherever the P.C.M.R. maintains a gathering place, and furnishing practical instruction in the use of Ranger weapons, field tactics, etc. Wherever they go they make a special point of arranging their training schedules to meet the working conditions of the community. If necessary, for instance, an instructor might spend four or five days at a logging camp or fishery in order to fit his instruction periods into times which would not conflict with the vital war production activities of the men.

Instructors are equipped with a set of small arms for demonstration purposes. Minimum armament for the Rangers will consist of rifles and sub-machine guns.

Deadly at Short Range

Most prized of Ranger weapons is the vicious little Sten carbine, or sub-machine gun. Though it weighs only six pounds, having been stripped down to the barest essentials, it has already inspired the men with a healthy respect for its lethal qualities in close action. While it may look like a toy from the 5-and-10, the Sten gun is no plaything. Fired from the hip, in a technique developed by the P.C.M.R. and since adopted by Ottawa, the Sten can be used with accuracy over a range of 50 yards, while shoulder-firing lifts the effective range to more than double this distance. The Sten bullet will kill at even greater distance, but at the sacrifice of accuracy.

Altogether, the unimposing looking Sten is proving an ideal weapon for such close range work as is most likely to be encountered in the heavily forested mountains of British Columbia, and the Ranger thus equipped has the comforting assurance that the pressure of his finger on the trigger will send bullets pumping into the enemy at the rate of eight per second. Mechanism can be adjusted for firing in bursts or singly to discharge the entire magazine of 32 shots with one pull of the trigger. In long-range work his 30/30 rifle is the Ranger's trusted and familiar standby, while a double-bitted axe comes in handy for infighting.

Ranger Training Camp

In addition to the training provided on the job, a Ranger Training Camp has been established at and is operated in conjunction with the Advanced Engineer Training Centre at Sardis.

Rangers go to this camp, not for routine army training, but for specialized instruction in such subjects as the use of weapons, map reading, use of military explosives in demolition work, field defences, bridge building and bush tactics, to name some of the salient features of the course. Instruction in demolition with military explosives is included because, while many of the men have had extensive experience in the use of blasting powder in the woods and mines and on railway and road construction crews and stump ranches, the technique involved in the handling of gun cotton and similar high-powered military explosives requires special training.

Perhaps most important of all the objectives aimed at in the Sardis Camp is the creation of a mutual understanding between the Rangers and the men of the Active Army. Here members of both branches learn that a community of interests and aims links their work and here the ground is laid for smooth and effective co-operation in action.

A valuable adjunct to the direct training provided by the P.C.M.R. is "The Ranger," an 8-page, semi-monthly training magazine, published at H.Q. under the direction of Capt. Kennelly.

This well-designed publication was created to supplement the instructions provided by the field men, and to carry information and training data to the remotest outposts. Devoted to articles on various phases of Ranger operations, illustrated with photographs and diagrams, its aim is to co-ordinate Ranger training generally and encourage the development of fighting methods best adapted to B.C.'s difficult and varied terrain. Typical of the material provided are the contents of Volume 1, No. 1, dated September 1, 1942, which include such titles as: The Sten Revives Old Art of Hip-Shooting; Aircraft Recognition; Realism in Training; Lessons From Crete; Visualizing the Ground from a Map; and How to Learn the Morse Code.

Vigilant Thunderbird

Two weapons familiar to most Rangers, a double-bitted axe and a 30/30 rifle, figure prominently in the design of the Ranger badge, together with the Canadian maple leaf and the appropriate motto, "Vigilans." Equally appropriate is the Thunderbird which tops the badge. For centuries the Thunderbird has spread its protecting wings above the villages of our Coastal Indians, capping the huge old cedar totem poles whose gaudily colored shafts are a striking feature of the coast scenery. To the student of "totemism," however, the P.C.M.R.'s own bird will be unique. Unlike the originals, which gaze straight out into space along either side of a massive, horizontally projecting beak, the Ranger's Thunderbird, obeying the exhortation of the motto over which he hovers, has his nose in the air and his eye cocked warily upon the heavens, from which nothing blessed is anticipated via the agency of Japan.

Horsesense and Initiative

As indicated in the opening paragraph, the muster roll of the Rangers is kaleidoscopic in character.

In selecting officer personnel, Colonel Taylor has taken into consideration the fact that the organization is one which places a premium on individual drive and resourcefulness. His men and their leaders must, above all else, be self-sufficient, ready to act on the dictates of their own common sense, and prepared to operate for indeterminate periods without the assistance of supporting services. They must be able to forage for themselves and they must know the country well enough to take advantage of every favourable physical feature. With these facts in mind, he has subordinated consideration of civilian status in favour of the personal qualities of horsesense and initiative, which together comprise the basic qualifications for leadership. Provided a man possesses these essentials, plus the backing of his community and an intimate knowledge of his territory, he is regarded as promising officer material, whether or not he has had previous military experience or special educational advantages.

Outstanding in the record of all the men who now go to make up the sprawling network of the P.C.M.R. units is the community of outdoor interests. Work may hold some of them behind desks, draughting boards or counters, but all, without exception, are outdoor men by practice and inclination. In the space provided in their application forms for the listing of hobbies appears almost every possible outdoor pursuit from hunting, fishing, skating, skiing and sailing, to mountain climbing and ornithology. Many of these men have held positions of responsibility in their districts for half a lifetime and are thoroughly familiar with both geographical characteristics of their territory and the distribution and composition of the population. Others have gained an intimate knowledge of the country and its peculiarities as loggers, prospectors and trappers, knowing the woods, the streams and the mountains as we know our city blocks. The lighthouse keeper and the fisherman would be among the first to resist a coastal attack, while in the interior of the province, the farmer, the rancher and the fruit grower keep a watchful eye on the valleys and homesteads.

These are the men who know British Columbia and who will probably be the first to know and act if an attack is launched upon any remote point. Like their counterparts of an earlier day, they will have the job of fighting delaying actions, of harassing the invader from ambush and of keeping the regular forces informed of enemy movements. Against them, some day, may come a foe who has already demonstrated himself a master of guerilla tactics.

It is in preparation for such a day that upwards of fifteen thousand Pacific Coast Militia Rangers are learning the gentle art of guerilla warfare—pooling ideas on bush fighting and putting them into practice the hard way in realistic interpretation of what could happen here.

– 4 –

British Columbia's Guerilla Army: Masters of Camouflage Are These Eyes and Ears of Fighting Forces

Peter Madison

***Vancouver Sun*, 23 January 1943**

Intensive instruction is received by these intrepid men at their training camp near Vedder Crossing, and these instructions embrace every essential phase of the grim work they may be called upon to perform.

They live in that wild part of the country where an average regular would get lost in an hour, and it's no place for the jeep or tank. The Pacific Coast Militia Rangers may not know the finer points of the rules, but they'll get 98 out of 100 firing a Sten from the hip on the move.

The treachery of Pearl Harbor was the reason for their conception. Under the leadership of Colonel T. A. H. Taylor, soldier, surveyor and timber cruiser, these 15,000 fishermen, trappers, hunters, loggers and farmers are now becoming experts in the art of guerilla warfare and making certain that it "won't happen here."

Serve Without Pay

Masters of camouflage, they are the eyes and ears of the regular army. Their job is to guard the coast and interior and they know every physical feature of the area they cover. Dressed in strange greenish uniforms that blend into the terrain of tangled forest, they move quickly and silently through undergrowth and disappear into the foliage. It would take months to compile the detailed geographical information and local knowledge that they have at their finger tips and can transfer to an ordnance map if necessary.

At their training camp near [Vedder] Crossing, men chosen from Ranger companies throughout this province receive extensive instruction in map reading and drawing. Here they acquire knowledge of different types of rifles, anti-tank weapons and submachine guns. Detailed courses are given in the improvisation of grenades, demolitions, ju-jitsu, signalling and reconnaissance work. They are taught how to co-operate flexibly with the regular army.

Although volunteers serving without pay, Rangers are military in scope and status. Since they must often work alone in isolated areas, emphasis is placed on initiative in thought and action. Drill is kept at a minimum.

Small Mobile Groups

For the sake of elasticity the ranks are formed into companies, detachments and further reduced into groups of eight to 10 men. Mobility is easier for small groups of men co-ordinating in bush country, than for companies.

An everyday scene at the training camp is realistic practice in demolition work. Tank traps are set in the bush, the wires made invisible in undergrowth. Rangers crawl away into the bush. A few minutes later there is a faint wisp of white smoke, followed almost immediately by a deafening explosion. A tank, railway track or bridge would have been demolished. The men are taught to handle grenades with five to three second interval between lighting and bursting.

After completing his course, the Ranger returns to his company to teach other Rangers what he has learned. Many companies have built their own rifle range in the bush. The facilities of these ranges can be judged from the action taken by the commander of an RCN vessel who, while anchored off a West Coast point, had occasion to meet the local Rangers. He was so impressed with the range that he gave the ship's crew full musketry practice there.

Realistic Practice

No "hothouse" soldiers these, as can be seen from a description of an overnight "trip" published in a recent issue of The Ranger, their official organ, I quote: "Knowing that one of the requirements of a good Ranger is to know his district thoroughly, the commander organized the trip with this in mind. Each Ranger carried a pack consisting of blankets and enough food for 24 hours and his rifle. Signal flags, lights, Sten guns, axes, compasses and first aid equipment were also taken. Much valuable knowledge of the terrain was gained. An unnamed lake was discovered as well as two swamps no one knew existed.

"During, the expedition, realistic rifle and Sten gun practice were carried out. The Rangers combined stalking, scouting and ambushing with this practice. Route sketches recorded the distance covered and the time taken in moving from one place to another, as well as compass bearings."

Are Swift to Act

The aim of the Rangers is to have every section of our mountainous interior and coastline covered by well organized experts in modern guerilla warfare. At a moment's notice these men will galvanize into action, swiftly moving to prearranged strategic locations, co-ordinating to harass and repel hostile forces in event of an attempted sea or air invasion. From these positions a small group of well-trained men could ambush or prevent further advance of enemy landing forces for days if necessary.

Even then the picture is not complete. Our coastline is long and sparsely populated. Under adverse naval conditions certain islands and inlets could provide seaplane and land bases for offensive action against the mainland.

"Shoot to Kill"

The enemy parachutist and shock trooper is a thoroughly trained killer, who will strain every nerve to achieve his objective. Upon landing, usually in isolated areas, he has no scruples about what he smashes or whom he tears to bleeding shreds to accomplish his mission. This is the cunning savage the Ranger is being trained to outsmart. Here no book rules serve. He must shoot to kill, there won't be any second chance.

He will be the counterpart of his Russian guerilla brother. Not the lawless cut-throat as popularly imagined who seizes pitchfork or shotgun in a desperate attempt to harass the enemy invader while the regular army takes the full brunt of battle, but a thoroughly trained master of his art. In fact, so impressed have our neighbors [the United States] across the line been with Ranger training methods that they are organizing a similar army, to be known as Minute Men.

– 5 –

"The" Rangers

Marion J. Angus

National Home Monthly, July 1943

The smooth, green waters of a stone-walled Inlet which for days have known no movement but the swift flight of wings and the wind riffling its emerald waters are silently disturbed as a Jap sub cautiously breaks water...

Fog banks shroud the Pacific as swift, sinister, grey shapes speed across the silent waters, draw up at an isolated beach head and small, slant-eyed, brown men step quickly ashore and are absorbed into the amorphous landscape...

Winging down from the Aleutians or in from an aeroplane-carrier miles out in the Pacific, troop-laden planes shear the blue skies above the Coast and Cascades ranges and roar on to their destination somewhere in the Cariboo where snowy puff-balls float slowly down through the golden air, bearing paratroops on their raiding descent...

Fantastic pictures, you say? Ones that couldn't possibly happen here -- "here" being British Columbia, the most vital spot in Canada at the present moment?

Perhaps they can't -- but these are the blueprints for campaigns which have proved only too successful in the past and, theoretically, should be equally successful in the future.

Of course, there were factors in those other campaigns which won't appear in the British Columbia picture but, on the other hand, there are elements in this far Western set-up which are uniquely a local problem.

In Malaya, for instance, the Japs were helped by local "fifth columnists," natives whose knowledge of Jungle terrain and steamy tropical rivers proved invaluable to the invading Nipponese. These same factors which proved so efficacious in the speedy overrunning of Malaya will act in inverse ratio if the yellow men try their stunts here.

Why?

Because, just waiting for an opportunity to show what men forearmed, forewarned and determined, can do when it comes to defending their homeland are upwards 15,000 of the hardiest, toughest "hombres" who ever toted a gun or rode the range – outside of a western best seller.

True, a few of them haven't toted a gun all their lives: some of them can't even ride a horse and in this last west they are not called "hombres."

They are called loggers, bushmen, high-riggers, filers, fishermen, miners, ranchers, homesteaders, retired sailors and "old sweats" in their civilian capacity and "The Rangers" when it comes to defense.

"The" Rangers? Yes, for short: officially they are known as the Pacific Coast Militia Rangers.

The Pacific Coast Militia Rangers are something quite new: their organization was officially authorized in March of 1942 and it was especially designed by Western men to meet Western needs.

The Rangers are unique for many reasons. To begin with there has been nothing at all similar in Canada since the days of Indian fighting when every householder and settler went about his plowing, sowing, harvesting and wood cutting with his gun beside him.

The nearest modern comparisons are the Home Guards of Britain or the units in Switzerland who, at the ringing of church bells, will drop everything they are doing and spring into action to defend the mountain passes of their homeland, which has never been invaded since the days of Hannibal.

However, though comparable to the British Home Guards in some respect, "The" Rangers are different in others. The very designation is expressive of the most vital distinction -- the Home Guards' function is "to guard" which signifies something more or less static, while the Rangers' is "to range"; which suggests a more or less mobile set-up.

In an area as large as British Columbia and with such a varied terrain and possessing so few people a static organization would be useless.

"The" Rangers are more than guards: definitely they will not stay in one place in their own area but will be extremely mobile. They will scout: they will act as "the eyes and ears" of the regular army and will take action on their own if need be, according to Col. T.A.H. Taylor, Officer Commanding the Pacific Coast Militia Rangers.

When danger became possible on the West Coast, British Columbians became increasingly alarmed and suggestions poured in from many quarters that the regular armed forces would welcome local help.

This fact soon became apparent to both military and local authorities. British Columbia is a land of uneven terrain … of hills and high mountains … of valleys and level prairies … of second growth timber almost jungle-like in its luxuriance … of matted salal, growing high as fourteen feet in some places, and dense undergrowth ... of streams and lakes … of secluded bays, winding channels and isolated beach heads.

It would take new troops, coming from other Provinces, a long time to become familiar with British Columbian terrain, but the men who for years have worked in the woods -- logging, trapping, forest ranging and farming -- know every inch of their particular districts.

Not only do they know it through their work but most of them in their spare time take busmen's holidays hunting and fishing.

Similarly men who have spent their lives on the coastal waters are acquainted with bays and shoals, inlets and reefs, which are a characteristic of British Columbian coastal waters and shorelines.

It was therefore inevitable that some sort of close co-operation between the regular military authorities and local authorities would be helpful. Even long before the war responsible citizens were aware of the need of some sort of flexible defense and plans were quietly made and preparations under way a long time before the Rangers were officially started.

Fortunately for the British Columbian people, the Honorable Minister of Defense, Col. J.L. Ralston, D.S.O., was sympathetic to the idea as well as appreciative of the need. When the time came this sympathy and appreciation resulted in the authorization of the Ranger organization which grew so rapidly and so far outstripped what was originally intended that it was difficult for departmental services and facilities to keep apace with the demands of this new force.

As well as the actual need from the viewpoint of military and local authorities, there was a keen desire on the part at British Columbian men to have such an organization and the idea spread like wild-fire.

Who can say how or when that desire was first conceived! Perhaps the first "vet" who offered his services -- nearly everyone of them who possibly could did -- at the outbreak of the war may have said to himself, "I was one of those who 'didn't finish the job' last time, but, by God, if they'll only put a rifle in my hands I'll do it this time, so help me!"

Perhaps the first woodsman, inured to the outdoors and all sorts of weather and hardship, who learned to his disgust that "because he was beyond the age limit" he couldn't join up … perhaps the first old campaigner who had fought on the Empire's frontiers or the first old marine or naval captain who learned to his sorrow that His Majesty's Government was grateful for past services and regretted it could not avail itself of present proffered services … most of all those who for one reason or another were precluded from joining the regular forces yet burned with zeal to defend their country in time of peril ... all these or anyone of upwards of 15,000 men who are now Rangers may have conceived and spread the idea of just what they could do "given the chance."

As the need for increased protection became more apparent after the bombing of Pearl Harbor, and as the desire of British Columbians became more vocal, not only in the exposed coastal areas and fishing banks, but in logging camps, isolated trap-lines, stump-ranches and strategic points in the interior, civilian authorities from the Provincial Government right down to municipal councils offered their co-operation.

Logging companies and loggers, especially, were most enthusiastic. One logging official remarked recently, "The building up of Ranger units should be a real factor in our defense and it has already done much to build up morale in the lonely districts surrounding our logging camps."

When the organization was first decided upon, the man who was placed in charge and given the job of designing and implementing an organization to suit the peculiar conditions met with on Canada's West Coast was Lieut.-Col. T.A.H. Taylor, O.B.E., M.C., a veteran of the last war. He was taken from an important staff post at Pacific Command Headquarters and given complete charge and responsibility.

Fortunately for the Rangers, Col. Taylor was not just an ordinary "brass hat," and *au fait* only with military organization and procedure. He had had experience in the bush: in lumber mills, logging camps and on the coastal waters of British Columbia.

In his younger days, he had engaged in land surveying, timber cruising and railway construction in some of the wildest parts of the Western Province, both coastal and interior. He knew the tricks and "quirks" of the country and the people in it, and he realized that if any organization was to succeed, it had to be designed especially for the purpose.

When Col. Taylor is complimented upon the rapid growth of the Pacific Coast Militia Rangers, or the Ranger movement as many call it, his invariable reply is, "It just grew up like Topsy." It is indigenous to the country and with the problems we may have to meet it could not have been handled otherwise.

With his first-hand knowledge of construction work both in northern British Columbia and on the Gold Coast of Africa, Col. Taylor is aware of the speed and skill with which logging crews and construction men can improvise and build bridges or demolish routes which would be of aid to the enemy.

"Our wild mountainous coastal areas offer most attractive possibilities to our enemies," said Col. Taylor. "Much of our terrain is well known to the Japanese through their fishing and logging activities. Only experienced men accustomed to rugged, timbered country could adequately undertake much of the work which would be required if the Nipponese gained a foothold and "The" Rangers are the ideal answer to this need. Strangely enough, the initiative and energy possessed by many of these men would not fit them for the role of an ordinary soldier where unified action is imperative."

"In coastal areas, locally known trails are often the only means of communication. Our woodsmen know these trails. In their own areas they could travel through fog or snow, something that no one unacquainted with the district could possibly do."

And his Ranger officers and men? Case histories reveal most interesting personalities and adventure-filled lives.

Here are a few, chosen at random from files and personal interviews:

The famous Col. "Cy." Peck, V.C., D.S.O., formerly officer commanding the 16th [Battalion], [Canadian Expeditionary Force], in the First Great War, known and beloved through British Columbia.

Lieut.-Col. A. Leslie Coote, whose fifty years in British Columbia have made him the best-known man in the Fraser Valley.

A fur trader with 15 years bush training, familiar with Indian (B.C. and prairie) dialects and northern transportation methods, whose hobby is amateur radio.

A man who served in the British navy for 18 years, his last position being chief skipper and who is now head net-man, of whom his sponsor reported, "The Ranger Captain in this area is a very responsible post and … is deemed the logical man in this community as any Ranger movements would involve the use of small sea-craft with which he is thoroughly familiar as well as being a gunnery expert."

A 35-year-old Coast Indian who is captain of a fish packer and knows the coast waters like a book … aggressive and reliable and has been councillor of his village for 10 years.

One could go on with saw-filer, baker, submarine chaser, caterpillar operator, game warden, fishery inspector, loftsmen, cowboys, farmers and a host of others, all vital, key men in this democratic army.

Col. Taylor, who is a colorful personality himself, is very broad-minded about his Ranger officers.

"I don't hold it for or against a man," he said, "that he has worn 'the old school tie' or served in previous campaigns. Equally I don't give a continental if a man has to make a cross for his signature. In fact, I don't care what a man has been or has done. The thing that concerns me is – "Is this man the logical Ranger captain for this district? Does he show qualities of leadership and initiative and a thorough knowledge of local conditions and terrain? If the answer is in the affirmative, he's the man I want." With this in mind, key men were selected in each district. A southern Vancouver Island area was the first district to form a Ranger unit. Other districts followed in rapid succession.

The beginning was modest -- no uniform, no insignia, no equipment. Then, on July 31, Col. Ralston told the House of Commons that weapons, including Tommy guns, had been sent to the Pacific Coast Militia Rangers and that other equipment would follow shortly. This gave impetus to recruiting. Men flocked to join existing units and other units were formed. The British Columbian men have something tangible.

A uniform has been designed, similar to the clothing worn by practical men who work in the outdoors. Even this is a "flexible" costume and can be worn to suit local conditions. It is made of special windproof and wetproof material.

The insignia was designed and here Col. Taylor chose something peculiarly British Columbian, the "Thunder-bird," which has long been identified with the heraldry of the Coast Indians.

In Indian art the "Thunder-bird" looks straight ahead but on the Ranger insignia its head is uplifted in an alert, listening attitude.

The organization of the Ranger units was also "flexible" with initial "key officers" but no set number of men. This is regulated by the needs of a district, the number of men available and diverse other things, and does not permit of a stiff, regulated pattern.

As the organization must fit local terrain and personnel offering, drill and manoeuvres are also elastic, depending solely upon conditions in a given area.

Ranger units take the distinctive names of their areas. For instance, on Vancouver Island, there are the Southern Vancouver Island Rangers, the Saanich Rangers, the Cowichan Rangers and the Alberni Rangers, some of the earlier units to be formed. These are in the front line of Island defense and fill vital roles.

They consist mainly of loggers and woodsmen, and fording streams, building or demolishing bridges and similar tactics might fall to their lot.

Fraser Valley district has units consisting of farmers and all types who make up a rural community.

Cariboo units are made up of cowboys, ranchers. One hilly district has a unit composed entirely of mounted men as horseback is the most suitable way of traversing the district.

Around the Chilcotin and Williams Lake districts, most of the men are ranchers who are natural "Rangers" as they know every inch of the country, and cowboys. Indicative of the way the Indians are backing the war effort was the 102-year-old Dog Creek Indian who offered his services as a guide or marksman and pointed back to a long, successful career as suitable qualifications. In appreciation of his sincerity, he was made an Honorary Ranger.

Nanaimo Rangers are mostly miners – "Geordies from Newcastles," Poles, eager to have a chance at any enemy who sets foot on these free shores, and Scandinavians.

Contrasting with the above are the West Point Grey and Hollyburn Rangers, composed chiefly of city men.

During the long summer evenings and on autumn Sundays, dignified business men can be seen crawling on their stomachs in a manner reminiscent of long ago boyhood days when they played Indian scouts. Not a quiver of a fern reveals their presence as they go through the woods. They shinny over waterfalls or ford streams with the elasticity of youth. In winter they will study map and compass-reading, first-aid, signalling, target practice and other branches of Ranger work.

I was privileged to see several Ranger manoeuvres and as they differ in details, short accounts will show the diversity of Ranger activities.

Coquitlam on the mainland was the first. This is a flat country with farms and dense second growth timber right at the roadsides.

It was dusk when we arrived at the level meeting ground. Immediately the Rangers split into two parties -- a group of seven being allotted the task of playing Jap parachutists who had ascended at a strategic point and escorted themselves there.

The main party march half a mile down the road.

"It takes a parachutist about fifteen minutes to recover from his descent, bury his parachute and get his equipment ready," the sergeant-major told me. "In that time our Rangers should be able to surround them and mop them up."

"How can they surround them?" I asked. "That's almost impenetrable growth on the other side of the road: they can't come up this side without being seen, and as for getting back of us…," I shrugged expressively at the impossibility.

"Military secret," grinned the [sergeant-major]. "However, I can tell you that for slashing through the undergrowth they have a specially designed bowie knife. Most of these men have lived here all their lives ... back in the days of the 'big timber.'"

I looked at the tall Douglas firs across the road. "Aren't those 'big timber'?" I asked.

"No. Second growth. With the rain we have on the coast the growth is almost jungle-like in density and luxuriance."

Zero hour approached. I strained my eyes, watching the ferns, three and four feet high, for the slightest movement. In the dusk they stayed motionless.

The mosquitoes began a private "blitz" on me.

"Now," I thought, "I'll see some of those Rangers. They'll simply have to scratch."

I just didn't know those Rangers. Their tough, tanned leathery hides were impervious to puny mosquito stabs when their owners' minds were concentrating on business.

Later I talked to some of the Rangers and asked if the mosquitoes didn't bother them. They laughed. One man said, "You should have been here three weeks ago. The mosquitoes were so thick," he made an expressive gesture down his arm, "we just brushed them off like this."

Zero hour came. From a distant comer of the field came the rattle of gun fire. We were in direct line of fire. The "Jap" nest was to our right and behind us -- and though we knew the shells were blank, involuntarily we cringed and stepped back into the bracken. Stepped, in fact, almost on top of a crouching Ranger. How he got there I'll never know. I had watched the spot intently, thinking the Rangers might try just such a stunt to fool me but not a bush nor a fern had trembled.

Things moved fast and furiously. Rifle-fire and machine-gun fire shattered the twilight stillness. Rangers crouched, advanced, sank motionlessly behind cover and advanced again from all directions. The "invaders" were completely surrounded and in a matter of moments almost wiped out. Mopping up took another minute or two … and a pleased look crossed the Ranger Captain's face as he looked at his watch.

"Not bad," he commented. "That's two minutes off last time and we'll better it. If…" He didn't need to finish the sentence and I didn't doubt he was right. Even in those mock manoeuvres the boys looked grimly business-like.

After it was over I asked one of the men, "Why have you joined the Rangers?"

"To defend my home," he said simply. "My home and my family." A minute later three small tots came running up and a childish treble piped, "Did you get the Japs, Daddy? Did you kill them?"

And on a Sunday I saw the Mount Arrowsmith Rangers at Parksville in action. Parksville is strategically situated at the crossroads where north and south highways meet the main cross island highway and an advance from the west would put the enemy in a good position.

Previously this attempt had been realistically enacted with the "Japs" sweeping in from the Alberni mountains with tanks, machine guns and other equipment. From the moment they left the mountains they were ambushed from north, south and west while a strong force advanced from the east to meet them. The "enemy" was hopelessly out-flanked and outmanoeuvred and at last annihilated.

This Sunday was different. A party of Japs -- the "White" army -- supposedly had landed from barges to the flats and worked their way inland towards the bluffs. Accompanied by one of the Ranger Lieutenants I went to find the "invaders." They were well concealed in a clump of young trees. We found them and stayed with them, following their cautious advance from tree to tree, from clump of bushes to bluff...

Then we pushed boldly ahead onto the highway. We were promptly halted by a guard and made to identify ourselves. Every motorist was halted and given instructions. At first most of them thought the whole thing a joke but as gunfire increased and smoke-barrages became heavier, thoughtfulness replaced amusement.

The defenders -- the "Red" army -- had not been caught napping. Every man was at his post: snipers on the roof-tops; boy runners on the job. The "Reds" had taken control of the telephone and telegraph services. The first Aid depot was manned and "casualties" began arriving via ambulance -- a delivery truck. Certain Rangers had been detailed for [Air Raid Precaution] duties and were standing by with hoses and fire-fighting and gas-decontamination equipment.

To those of us who had never been "under fire" it was a strange, rather terrifying experience. A wind had arisen. The smoke-screen eddied and swirled. Through it, misty shapes armed with Tommy guns and hand grenades could be seen vaguely.

The "invaders" had aerial support and "thunder-flashes" represented falling bombs. For the occasion the ammunition had been supplied by the military authorities. It was probably the largest amount of ammunition ever expended in that part of the world.

A realistic "incident" happened when a "Red" staff car drove up to the telephone office. From the trunk of the car emerged a "White" soldier who shot the officer in charge with out parley. He, in turn, was shot by an unseen sniper and the "Reds" remained in possession.

We went to the First Aid Depot. There was a business-like set-up with mattresses, uniformed nurses, bandages, hot water bottles and other paraphernalia. "Casualties" in various degrees of distress completed the picture. The realism was intensified when word came the "Whites" had surrounded the depot. Our trusty defenders were on the job, however, and the attackers were repelled.

There was one tragi-comic incident. A black spaniel, curious and nervous between the thunderous noises, bright flashes and strange sulphurous smells, decided to investigate. Result -- he arrived just as a rocket went off and a more astonished pup never scampered back to his owner.

When the bugler sounded, "Cease firing!" and casualties, tactics and results were toted up, the referees decided the defending army had successfully resisted the invaders -- a happy token.

The following morning I saw two of the five rifle ranges in this general area. The Rangers have 30-30 Winchester carbines, which are more suited to bush and range fighting than the military rifles. One of these is on the flats and has a 100 yd., 200 yd., and 1,000 yard range. It was built with difficulty as the sea water kept seeping through and dislodged the general contour. This has been rectified and improved.

The other was deep in the woods, a grim reminder of the imminence of danger contrasting strangely with the cathedral stillness of the giant Douglas firs.

The story of this range is one of the many examples of Ranger ingenuity. From an abandoned logging trail, one branches off onto a good road made by the Rangers themselves. All the excavating, banking up of the butt embankment, the felling of the huge trees and placing them firmly interlaced into position was the work of the men in their spare time.

The recording trench is 32x8x8 ft. with a two-foot overlap and is a marvellous sample of bush construction work.

Suddenly I said to the Ranger Captain, "Wouldn't it be a strange thing" if the Japs did come and, like wraiths in the cool green twilight of our northern woods, would come forth the avengers of Hong Kong… Singapore… Bataan and New Guinea?"

And he answered grimly, "Strange, yes … but as logical as anything in this war. If that time ever comes we will be waiting: more, we will be ready."

Personally I agree from observation and conversations with Ranger Captains and their men. Eight trained field supervisors periodically visit units and maintain contact with headquarters and operational commands. Travelling instructors with equipment are provided. They visit outlying units and give instructions to the men after working hours. Selected members receive instructions at a central training camp where courses are continuous and hundreds will be trained yearly and returned to pass on their knowledge to fellow-Rangers.

The Thunder-bird is on the watch guarding its immemorial home -- British Columbia's coastal areas. His cohorts are some fifteen thousand Rangers who will -- when the time comes -- prove worthy of the best traditions of the Empire's fighting men.

– 6 –

Vigilance is Their Motto

Jack Strickland

Vancouver Daily Province Saturday Magazine, 7 April 1945

From the Arctic Ocean to the southern border, from the Pacific Ocean to the Rockies, British Columbia's modern "Vigilantes" -- her 15,000 Pacific Coast Militia Rangers--look westward to dangers that may come, and laugh them off with high-felt confidence.

Security veils the full extent of their defense preparations, which take in a jagged coastline of 5000 miles, which cover and area of over a quarter million square miles of some of the roughest, wildcat terrain in the world.

"Patriotism? Rubbish," says their commanding officer, Lt.-Col. Thomas A.H. Taylor, O.B.E., M.C., veteran soldier and outdoor man.

"These men are banded into a close-knit body with but a single purpose--actual defense of their own homes. They know their own country inside out, its every thicket and bush and rock, its very deer tracks."

High-ranking officers who have studied and observed the Rangers at their work, go further.

They believe that had the Ranger form of outfit--organized local defense units-- existed, not only in B.C., but in other parts of Canada, when war broke out, there would be few army deserters, few "zombies."

From 15 to 70 They Foregather

As one officer put it: "When an army recruiting truck wheels into Quebec loaded with strangers and a loudspeaker blares, "Come on pal," Jean Baptiste just doesn't get it. Let him organize his own "Courieurs de Bois" and you've got a springboard for recruiting that's a 'natural.'"

They'd like to see outfits like the Rangers in every part of Canada--formed from men like the men of B.C., from 15 to 70, where discipline is not autocratic, where "esprit de corps" is the rule. Leaders of the group, sometimes called by outside observers, "the oddest army on the continent," are natural leaders in their own communities.

There are no outsiders. There are no "parade ground soldiers." "Vigilance" is their simple motto.

They have done work that no other outfit in Canada would even attempt. Civilians, serving without pay, they have been scouts and guides for the Active Army, and should invasion come, would be its auxiliary fighters.

Whenever a plane is down in the trackless wastelands of B.C., the Rangers are there to comb and re-comb every inch of brushland.

Recently, a hunter disappeared in wild bush country on Vancouver Island. A thorough air search scoured the countryside; but still the boy's parents were not satisfied.

The Rangers went to work. They covered a tract of 64 square miles, with trailwise Rangers searching the entire area, spaced only five feet apart.

The hunter was not found, but the parents were certain, then, that he was not there.

There will be no "stab-in-the-back" though Canada's northland back door. White clad Rangers in the frozen Yukon are on 24-hour alert with their dog teams, their sleds, their rifles and Sten guns.

There will be no further sneaking in from off-shore. Like Britain's famed "little-boat" miracle men at Dunkerque [Dunkirk] the Rangers have their men with boats.

Its marine section is a fleet of boats manned by their owner-Rangers, sea-wise, sea-ready, and ever on the watch.

For Defense of Rural Districts

In the interior reaches of B.C. Ranger companies are spread over the land--companies of fully-trained "serious soldiers," as their officers call them, who can lead a regular soldier through their home territory and lose him at will, and who can shoot straight and have the ammunition to do so. Caches of food and ammunition are hidden and ready for emergencies, for the Ranger is a practical man.

He does not clash with his fighting brother in the Reserve Army, for their jobs compliment, rather than contradict each other.

The Reserve Army units are built for defense of metropolitan areas, the Ranger units for defense of the rural districts. Rangers work in territory where the Reserves would be use-less, the Reserve men in areas where the trailwise Ranger would be lost.

Ranger ranks are made up of men from all walks of life--loggers, fishermen, hunters, businessmen. They meet in community halls, in Legion halls and church basements. They range in age from 15 to 75.

The P.C.M.R.'s were born out of the treachery of Pearl Harbor. They were authorized by Ottawa on March 14, 1942, following a series of public requests for organized local defense units, and receipt of a deluge of letters by military and civil authorities.

Within four months, over 10,000 men had swarmed into over a hundred B.C. centres to enlist.

Units are part of the Canadian Army, not unauthorized "guerilla bands." Enlistment does not interfere in any way with civilian work.

Their small headquarters staff in Vancouver Barracks of Pacific Command is about their only expense.

Organized to Combat Forest Fire

By August 1, 1942, Rangers in some coastal sections were practicing with new Sten sub-machine carbines and were using army "pull-throughs" on service style rifles, but very few units got a full slate of weapons that year.

So machinists of the Ranger ranks made weapons in their spare time; electricians improvised signalling equipment; chemists mixed "Molotov Cocktails" from explosives and empty beer bottles; ballistic experts made up ball ammunition and tracer bullets.

Many made detailed maps of their areas during weekends; others checked on roads and bridges to supply military authorities with "engineer intelligence" of their districts.

When war came there were five rifle ranges in B.C. The Rangers built 163 more, at no expense to the public.

Companies have trained their own homing pigeon messengers. They have their own training school for instructors, their own training officers and travelling instructors.

They know and love their weapons--rifles, carbines, sub-machine guns, Sten sub-machine guns and 30-30 sporting carbines.

They learn fieldcraft and reconnaissance work, and out of their skill and intimate knowledge of local conditions, the army has found many of the training sites it uses.

Forestry officials, worried by the ever-present menace of fire in B.C.'s vast forestlands, have found a deep relief, thanks to the Rangers. They are organized to help in keeping forest fires under control.

The Rangers have their own monthly magazine, The Ranger, a high-grade paper job devoted chiefly to training subjects.

– 7 –

Tribute Paid Rangers At 'Stand Down'

Tony Skae

Vancouver News-Herald, 1 October 1945

Another chapter in the history of Canada's role in World War II was brought to a close Sunday when the official "stand down" of the Pacific Coast Rangers was held at Brockton Point Oval.

For the last time, approximately 1,500 Rangers, garbed in their customary "dry-bak" hunting clothes, and with rifles at the sling, marched briskly past the reviewing stand while hundreds of spectators watched Lt.-Gov. W. C. Woodward take the salute.

Piped into the field by the band of the 2nd Battalion Canadian Scottish Regiment and marching in review order to the band of Little Mountain Barracks, tribute was paid the disbanding Rangers by a company of Canada's active army, the King's Own Rifles of Canada.

Gratitude was expressed by Lt.-Gov. Woodward and Lt.-Gen. F. F. Worthington, C.B., M.C., M.M., [General Officer Commanding-in-Chief], Pacific Command, to the men who, for the past three and a half years, voluntarily and without remuneration, defended some 5,000 miles of Canada's Pacific coast.

Owed Great Debt

To the Rangers "who provided almost the only defence on the coast" during the early years of war, when the Pacific coast was almost without protection, Canada owes a great debt of gratitude, Mr. Woodward said.

Gen. Worthington praised the co-operation extended by the Rangers to army and air force, especially to their vigilance in discovering and aiding in destruction of the Japanese balloons.

Although probably only one person in a thousand in Canada knows of their existence, the general said, the Rangers gave "the greatest possible service" to Pacific Command before Pacific defences were set up.

Personal thanks from the general were extended the wives of the Rangers who, while on the "only unrequited service to the country," were often absent from their homes for many days.

Expert Guerillas

Authorized for service in February, 1942, the Rangers had reached some 10,000 strong within a few months of organizing.

Fashioned after the "Partisans" of Russia, the Rangers adopted guerilla warfare and included in their training map-reading, reconnaissance, field sketching and map drawing. Many unmapped regions of British Columbia were mapped by the Ranger men.

With them bushcraft was a highly developed art and it was disclosed by the general that on many occasions the Rangers directed active service troops to alternate routes of evacuation during mock battles.

Recruited from as far afield as the Yukon and Peace River, to the Kootenays and Vancouver Island, the Rangers patrolled an area as large as the combined areas of Belgium, Switzerland, France and Italy.

Attracting little recognition and with no glamor attached to their work, the "stand down" Sunday represented Canada's "thanks" for a job well done.

– 8 –

Claxton Announces Formation of New Ranger Organization

Vancouver News-Herald, 9 April 1947

Organization of a Dominion-wide "people's" defense force made up of trappers, miners, woodsmen and farmers in remote parts of Canada as an addition to the reserve Militia was announced here Tuesday by Defense Minister Brooke Claxton at the conclusion of his six-day inspection of West Coast defenses.

The organization will be patterned along the lines of the Pacific Coast Militia Rangers, wartime civilian organization which figured largely in defense of lonely outposts along British Columbia's coast.

The defense minister pointed out formation of this new defense unit made possible the revival of the Rangers, which were disbanded in 1945.

"The Canadian Rangers (official name given to the new corps) will be formed on a restricted basis across Canada for the purpose of operating in thinly populated parts of the country, which are not normally traversed nor under surveillance and where it is impractical to organize units of the Reserve Force," said the minister.

Their role in peace includes acquiring and reporting detailed information about the country, likely to be of use in future planning.

"The possible wartime role is envisaged as that of guides, reconnaissance, watchers and co-operating with the defense forces, RCMP and Provincial Police in various ways," he said.

The present plan, said Mr. Claxton, is to form nucleus companies of key personnel capable of rapid expansion. Men will be formed into companies with subunits which will be highly flexible in organization to allow for their varied duties.

No establishment will be laid down by the defense department as to the size of the organization and personnel in it will not come under military discipline.

Issued Rifles

Personnel will be issued with rifles by the army but will not receive pay unless called on for specific service. Training will be left largely to the units themselves.

In making the announcement, the minister paid tribute to the PCMR organization for its wartime work of guarding thinly-populated areas against possible Japanese attacks.

He made the announcement following conferences with Lt.-Col. T. A. H. Taylor, former head of the Rangers and present managing director of an association formed to perpetuate the unit on a civilian basis.

Maj.-Gen. F. F. Worthington, commander of the army's strategic Western Command, participated in talks between Mr. Claxton and Col. Taylor.

Mr. Claxton indicated Gen. Worthington would be given a supervisory role in formation of the organization in the west. He has been closely linked with the formation for such a civilian organization as part of Canada's defense forces and has repeatedly advocated such a plan.

Minimum Force

Asked if he felt the present size of Canada's armed forces was adequate for the country's defense, Mr. Claxton told newsmen the aim was a minimum force in keeping with the country's need and he felt that it would be "adequately developed to do that job."

"I've never heard of any fire or police force or any other kind of a force that was said to be adequate," he added.

– 9 –

First Ranger Group Formed at Dawson

Vancouver News-Herald, 7 October 1947

Dawson, Y.T. — No. 1 Company of the newly-conceived corps of Canadian Rangers is now in the process of being organized in Dawson City, Yukon.

It is reported that similar companies will be organized throughout the Yukon and Northwest Territories.

Establishment of the first unit in Dawson is considered a signal honor for this far northern mining centre.

Plan Outposts

Outposts of No. 1 Company will be established at Stewart City, Y.T., Mayo and Selkirk. No. 2 Company will be organized at Whitehorse.

Initial steps to get the new organization under way were taken when a small group of Dawson men, including war veterans and former members of the now-disbanded Pacific Coast [Militia] Rangers, met Brigadier-General G. A. Walsh, CBE, DSO, officer commanding the Northern Highway Systems.

Chapman Chosen

At the initial meeting C. H. Chapman, former commanding officer of No. 135 Dawson company of the Pacific Coast [Militia] Rangers, was proposed as the chief officer for the new unit.

According to Mr. Chapman, former member of the RCMP, but now one of Dawson's leading businessmen, the new company will be composed of older, experienced men, capable of acting as guides in any part of the territory. Few details have as yet been made known regarding the exact duties and purposes of the new organization, but it is understood that it will be well equipped. Army officers will be making regular trips to the north to inspect and instruct members of the Yukon companies.

Congratulations

When Major-General F. F. Worthington, C.B., M.C., M.M., since retired as general officer commanding, western command, was advised that Mr. Chapman had been named to head the company he sent the following congratulatory message:

"My hearty congratulations at your being the first officer appointed and in command of the First Ranger Company in Canada's newest defense organization. It is also gratifying to me that an old RCMP officer leads the van and I am confident No. 1 Dawson City Company will be first in many other things under your able guidance and direction."

– 10 –

Canada North Defence is Set: 'Ranger' Units as Home Guard to Include Hunters, Trappers

Montreal Gazette, 20 March 1948

Ottawa— A sprawling network of trappers, hunters, farmers and lumbermen will be Canada's first line of a Northland defence in event of a military emergency.

Organized into "ranger" units—a Far North home-guard of minute-men — they will have the responsibility for the immediate defence of vital points.

Armed with rifles and probably Bren guns, tommy guns, grenades and pistols, they will stand guard [over] towns, mines, roads and airfields until regular combat troops are sent north, if developments warranted.

These Arctic guerrillas, with intimate knowledge of their own areas, are a comparatively new development in Canada's Arctic defence plans.

Military authorities here scoffed at suggestions that Canada's lone combat brigade of about 3,000 men would be immediately thrown into the Northland, even if war with Russia broke out.

These troops form the narrow backbone of the army and if mobilization came they would be the foundation on which the wartime army would be built.

One point that would be protected quickly, however, with some active troops backing up the Rangers would be the Eldorado Mines in Great Slave Lake in the Northwest Territories, where uranium is obtained for atomic development.

The vital communication route of the Northwest highway from Edmonton to Whitehorse and on to Alaska would also be patrolled and guarded.

A senior army officer said that Ranger units are being organized across the Northland in co-operation with the R.C.M.P., the Hudson's Bay Company and the two Canadian railways.

One company is located at Whitehorse and has been trained by the Royal Canadian Engineers working on the Northwest highway.

The Rangers are not members of the active army. They are drawn from the citizens of towns and hamlets and the open spaces of the Northland and have approximately the same status as members of the Reserve Force.

– 11 –

Know Every Creek, River, Hill
Canada's Rangers 'Eyes of the North'

Robert Nielsen

***Toronto Daily Star*, 12 May 1948**

Ottawa—The defence department assures the taxpayer he's getting his money's worth out of the Canadian Rangers. A spokesman says they are useful even in peacetime, and will be practically indispensable in the event of war.

The Rangers are a body of woodsmen, trappers and guides being organized to maintain a lookout patrol in remote and unsettled areas of the north.

The defence spokesman said the number of Rangers can't be revealed, but it would be a "few thousand." Neither could he give the cost of training and equipment.

He listed four functions:

(1) As guides to troops who might be serving in the area.

(2) Preparing local defence schemes.

(3) Providing the army detailed information on their own areas. A ranger at Fort Nelson, for example, would be expected to know every river, creek and hill for miles around.

(4) As rescue parties in peace and war.

The army borrowed the idea of a rangers corps from the Pacific coast [militia] rangers, who kept watch for possible Japanese penetration during the recent war. They are not a part of either the active or reserve forces of the army, and get no pay unless called up for specific duties.

To add distinctiveness, the army is issuing an armband for the Rangers, bearing a green shield on which the letters "Canadian Rangers" are embroidered in red. The shield also bears three red maple leaves, superimposed over a crossed rifle and a double-bitted axe.

– 12 –

Unsung Ranger Arctic Warfare Defenders in 'Guerilla Warfare' for Army

Montreal Gazette, 13 February 1954

Silver Creek, YT—The Canadian Militia Rangers – North America's first line for Arctic defence – are demonstrating for the Army the guerrilla tactics they would use if an enemy ever invaded sub-Arctic Canada from the north.

A handful of the Rangers is taking part in Exercise Hot Dog II, a winter test of men and equipment involving two companies of the Princess Patricia's Canadian Light Infantry in these frozen wastelands 1,400 miles northwest of Edmonton.

The Rangers are staging repeated hit-and-run patrols on 128 paratroopers attempting to destroy a hypothetical enemy wireless station in the snow-bound mountains near Kluane Lake.

The five-day attack on the 10-man wireless unit, the first mobile striking force operation of the exercise, is scheduled to be completed Sunday.

The Rangers participating in the exercise are just a small segment of a little-known reserve militia recruiting throughout the Canadian north and the Arctic archipelago.

Their positions and numbers still are secret. It is known, however, that they extend from the Yukon to Labrador and are rapidly being organized in the isolated northern sections of the provinces. In the event of hostilities, they would immediately go on active duty.

The militia's first workout with the Army came last year during Exercise Bulldog near Norman Wells, N.W.T., involving the PPCLI and the Royal 22nd Regiment.

The only military equipment doled out to the Rangers – all of them experienced outdoorsmen – consists of a rifle and an undisclosed amount of ammunition, an identifying arm band, and one of the khaki berets discarded by the Army following the Second World War.

They receive no pay except for participation in northern army exercises, and attend no parades. They are organized in companies, platoons and section, and have regular army ranks extending up to captain.

The director and organizer of these northern guerrillas is Maj. C.B. van Straubenzee, a slim rancher and well-known outdoorsman from Enderby, in British Columbia's Okanagan Valley.

Once a lieutenant-colonel in a tank unit of the Three Rivers Regiment, he now has been called up from reserve ranks to active service and will be one of the Duke of Edinburgh's guides during his tour through the Northwest Territories this summer.

The Rangers have grown gradually out of the former Pacific Coast [Militia] Rangers organized during the Second World War to watch wartime activity along British Columbia's coastline. They were recruited and founded by Maj-Gen F.F. Worthington, now Canada's federal civil defence co-ordinator.

Their ranks are drawn from ex-RCMP officers who have settled in the north, trappers, guides, prospectors, Hudson's Bay Co. trading personnel, and a few Eskimos and Indians.

Maj. van Straubenzee said they will be accepted in Ranger ranks "as long as they can read and write and count up to 10 – they don't necessarily have to be medically fit according to army standards." Each is a crack shot.

Should an enemy ever advance over the Arctic barrens, the Ranger role would be hit-and-run operations to stall the invading force until Canada's mobile striking force could be transported or parachuted into the area.

Little has been – or can be – said of their part in the northern exercise. Col. J.R. Cameron of Halifax, director of Hot Dog II, commented only that the Rangers were "most enthused."

– 13 –

Arctic Exercise: Indians Point the Way

Canadian Press

***Ottawa Citizen*, 22 February 1954**

Sept Iles, Que.—Indian Ranger scouts today led Canadian defence forces into action against theoretical invaders in the vital Sept Iles area as exercise Loup Garou entered its fourth day.

The scouts pointed the way to paratroop forces striving to knock out a tightly-moulded invading force represented by the first battalion, Royal Canadian Regiment from London, Ont.

They are the remnants of a local Canadian Rangers outfit made up of Indians, woodsmen and trappers who now are emerging as the heroes of this fast-paced Arctic exercise.

For days before the paratroopers hurtled into the Sept Iles salient to box in the northern attacking force, Rangers scoured the ground dodging the invaders but closely watching their movements as they set up defence positions along the fringes of Sept Iles airport.

Airport Captured

Only lack of radio facilities prevented the Rangers from channeling information to the defending air and ground forces building up strength at the huge Goose Bay airbase in Labrador. The invaders took over the airport Thursday afternoon with resistance as the Rangers, a civilian commando outfit, faded into the surrounding woods as they had been told to do if the enemy struck at this strategic iron-ore port and railhead.

They bided their time. Sunday they struck only a few hours before the battle-hardened Royal 22nd Regiment and supporting arms were parachuted into the area in a surprise attack.

The Rangers knocked out one of the enemy's two medium machine-gun posts, shot up a light machine gun outfit and captured a mortar position "with the loss of only one man."

Early Sunday, as the C-119 flying boxcars of the counter-attacking force followed in the bombers to drop their load of men and equipment by parachute, the Rangers emerged from their bush hideouts, donned red identification arm-bands and linked up with the airborne troops.

The counter-attackers struck in three waves, the first shortly after 6:30 a.m. AST., bringing in the initial drop of men and equipment, the second-called a re-supply, sending additional ammunition, food, sleds and other Arctic equipment floating earthwards, and the third, laid down at 1:40 p.m., placing battalion commander Lt.-Col. Trudeau and his headquarters alongside the rest of the airborne assault force.

Enemy Concealed

With weather closing in and rain threatening, the assault commander decided to push on and harass the invaders. But although the alien forces "killed" seven of his men in the first few minutes after the initial drop, the enemy withdrew to concealed positions which the airborne troops failed to find.

During the night Col. Trudeau sent out probing patrols aided by Indian Rangers. The rest of the local force was "wiped out" when they ran into an enemy ambush while on patrol. Night operations were restricted by the Indians' inability to read maps.

At dawn, the airborne troops prepared to sound out enemy defences between them and their objective—the airfield three miles to the south on the town borders.

The drop Sunday nearly ended in disaster when the first "stick" of paratroopers was jettisoned above timber and floated into tree-tops on the fringe of the landing area.

– 14 –

Rangers of Frobisher

Ambrose Shea

The Beaver, Winter 1956

Extracts from the diary of a Ranger [liaison] officer who, when three VIP's landed at Frobisher Bay, had to train an Eskimo guard of honour in less time than it takes to say itsuarkattarpok ["One keeps going out to see if someone is coming"].

Frobisher, N.W.T.

...

6 April [1956]

The [Commanding Officer (CO)] of the RCAF base has very kindly found me accommodation in the officers' quarters. This morning he and his opposite number, the American CO, ganged up on me, obviously by prearrangement, and suggested that the Rangers should provide a Guard of Honour for U.S. Secretary of Defence [Charles] Wilson; Hon. Ralph Campney, Minister of National Defence; Hon. C. D. Howe, and other VIPs who are due to arrive tomorrow. I am in favour of the Guard and think it would be very good for the Rangers but I would have liked more time for preparation.

With the help of Doug Wilkinson, Bob Griffiths and Sgt. Sageakdok, 14 Rangers were assembled in the townsite garage tonight, and with Sgt. Simonee acting as interpreter I explained the situation to them in such terms as I thought they would understand, then "sized" them and taught them to stand-at-ease and come to attention. I attempted no more than this and they caught on quickly in spite of the language difficulty. I have great hopes for them. It is to be a purely Eskimo show with Sageakdok as Guard Commander and Simonee as Sgt. I shall not appear at all but Bob Griffiths will stand by to provide any explanations that are needed.

7 April

Two plane-loads of VIPs arrived at 1145 this morning. There was some doubt as to whether they would arrive at all because of the weather, and when they did, they came in an unexpected order, the first plane bringing Hon. Ralph Campney and Hon. C. D. Howe, Canada's Minister of Trade and Commerce, instead of Mr. Wilson.

By this time. I had fastened the Guard in position and disappeared into the hangar behind them from where I would watch proceedings, more or less, through a crack in the door. When Mr. Howe alighted he incautiously approached within range of Sageakdok who at once called the Guard to attention as he had been told to do. The Reception Committee, after making some unsuccessful efforts to get him to reverse his decision, decided to bow to the inevitable and Mr. Howe inspected the Guard with which he was obviously fascinated.

I was fascinated also. None of the men concerned had ever heard of a Guard of Honour or done any drill until last night. They were dressed in their best clothes and for the sake of uniformity wore the hoods of their parkas up. Normally, Eskimos tend to slouch, but I had told them that soldiers were important people and that they should hold their heads high and not move a muscle while they were being inspected. They did this and were amazingly steady. As long as they didn't move anyone would have thought they had had months of training.*

Sageakdok, I knew, was nervous. He does not speak much English and was having difficulty in remembering the half-understood words of command and even in pronouncing them. Neither he nor Simonee, who speaks English well, can say "Shun." The nearest they can get to it is "S'un" and they were a little self-conscious about this. Furthermore, Sageakdok had to learn how to salute approximately correctly instead of in the sloppy style he has hitherto affected. In addition, until the inspection got under way and he had Simonee's support, such responsibility as there was rested squarely on his shoulders, which was probably a novel and rather frightening experience.

I need not have worried about him. He behaved with all the aplomb of a veteran [non-commissioned officer (NCO)], and I was both amused and amazed to see him stop and adjust one man's armband as he walked behind the inspecting party, looking each man over from head to foot as though he had been doing it for years!

I could not think where he had learnt to behave like this until it dawned on me that Eskimos are very observant and imitative, and that he was simply mimicking my own demonstration of the night before. It was an extraordinarily good performance and was only marred by the fact that he was, as I later learned, chewing bubble-gum throughout! He is an impressive-looking little man in a tough, compact sort of way and looked just right as Guard Commander.

Simonee, the section Sgt., is newly recruited and is very definitely an acquisition. I doubt if it would have been possible to mount a guard without his help. He is a young man, tall for an Eskimo, good-looking and very intelligent. He was one of the Eskimos who represented Baffin Island at the Coronation and he has the Coronation Medal, which he was wearing today.

He is at present employed by the Dept. of Northern Affairs at the new town site and is interpreter, carpenter and mechanic and will almost certainly be Mayor of Frobisher some day!

Mr. Howe had barely finished his inspection when Mr. Wilson disembarked and the whole thing began all over again. The Guard stood it nobly; in fact they

* Eskimos get used to this sort of thing when they stand motionless for hours over a seal's breathing hole.—Editor

were still standing rigidly at attention when I emerged from my hiding-place 15 minutes later and took the spell off them. They had not moved so much as an eyeball in half-an-hour.

I congratulated them heartily and meant every word of it. As far as I could judge, they were pleased with themselves. I understand that some of them met the Governor General when he visited here recently, but this is the first time that a formal Guard of Honour has been mounted.

– 15 –

Eyes and Ears of the North

Robert Taylor

The Star Weekly Magazine, 22 December 1956

Mile 1016, Alaska Highway —Three armed bandits fleeing down the lonely stretches of the Alaska Highway through the Yukon and toward freedom in the south had never heard of the Canadian Rangers. Yet, from the time they hit Canadian territory, there were under constant check by the Rangers. When they stopped at the Airport lodge at Mile 1095 they acted peacefully, so the watching Rangers didn't act.

Down south at Haines Junction the authorities were getting steady reports on their progress. At Mile 1016 Lt. Wally Wandga, Ranger platoon commander and camp foreman, mustered 10 Rangers and they blocked the highway with two road graders.

The bandits, in the second car they had stolen, were racing through the night. They screeched to a halt as lights flared, showing them the roadblock. They saw 10 rifles trained on them and quietly surrendered to the lone Mountie and his band of Rangers.

The affair was over in no time, but down in Ottawa the men in charge of the nation's security read the reports with satisfaction. Operation Bandits had demonstrated fast, effective co-ordination between the Rangers and the civil authorities, just the kind of help the military men hoped they would get when they set up the Rangers. They hadn't been thinking of bandits, of course: they'd been thinking of enemy agents, but the bandits provided a test the Rangers passed with flying colors.

The Rangers are, undoubtedly, one of the strangest military outfits in history. But they are tailor-made for one of the strangest military services in history—keeping watch over hundreds of thousands of square miles of the barren lands and bush country of northern Canada where the population averages one person every 80 square miles.

These men, trappers, prospectors, miners, woodsmen and farmers, are the eyes and ears of the defence department in 50 per cent of Canada, through the northern parts of the provinces and in the Yukon and the Northwest Territories.

Some of them can't read their own names but they are the real scholars of this country when it comes to reading signs on the trails of the north.

Eskimos, Indians, whites, and all the mixtures of these races, they are united in one task: Guarding a country that doesn't even know of their existence.

They watch and wait for any suspicious activities in their remote areas; they act as air observers to supplement the RCAF's ground observer corps, and they have provided a steady flow of reports that have demonstrated they take their duties seriously.

"We're Canada's first line of Arctic defence," says Capt. Dave Jones, of the Rangers at Aklavik. He speaks simply, yet proudly, conscious that he is a member of a carefully picked band of about 3,500 volunteers who are serving their country.

As well, they are the least expensive military force any nation has today. They don't get, or want, a nickel for their daily eyes-open work. Their only uniform is an army-pattern waist belt and an armband, a green shield bearing a crossed rifle and a double-bitted axe in black, over which are imposed three red maple leaves. In addition, each man is given one of the deadly 303's and an annual supply of ammunition for the rifle.

To get the same kind of protective screen of watchers from the regular army over this vast stretch of Canada would cost tens of millions of dollars a year.

However, army men, who would have to be changed at regular intervals, could never master the area as have these men who have spent their lives learning how to live with the north.

Formal army exercises in the north have shown that in the Rangers the regular military forces have the finest guides and scouts they could require. As commanding officer on one of these exercises later reported to Ottawa, "The Rangers not only acted as guides but also carried out active patrolling with excellent results."

For jobs like these they get the regular military pay of their ranks, since for such service they must leave their regular jobs.

Canadian military men quickly realized that in these men they have a strange sort of faculty of the college of northern knowledge. Stimulated by the interest shown at the various command headquarters, the Rangers work steadily to compile up-to-date data on their areas which is of considerable value to the regular defence forces.

In addition, army officers turn to them to learn about the country. Capt. A.J. Shea of the Royal Canadian Dragoons, Ranger liaison officer for Eastern Command, went up to Baffin Island for a sort of Arctic indoctrination course from the Ranger old-timers there.

Through Ranger Officer Robert Griffins, who is also the Hudson's Bay Co. post manager, and RCMP Constable Art Deer, he got trail-training and experience in living off the country in January that he has passed along in reports made available to all military districts in the country.

In one igloo he visited he found and old fashioned stone lamp in use, the first he had ever seen outside a museum. "Every now and again in this country one gets the eerie feeling of having travelled thousands of years back in time which is disconcerting—almost frightening," he wrote in his diary. "I return to the 20th century tomorrow; and I don't know whether I am glad or sorry."

Trips like these have convinced the military authorities that the Rangers would be invaluable if it were ever necessary to wage guerilla warfare in the northernmost parts of this country.

At Port Radium where Canada's rich uranium mine, Eldorado, is located, the cliff-like ramparts will be manned by a sharp-shooting company of Rangers if ever any effort is made in time of war or peace to raid that important plant.

Unsung, almost unknown, this northern home guard mans the distant, silent frontier country, experiencing through their service a new kinship with the men and women of the cities and farms of the Canadian south.

– 16 –

Shadow Army of the North

Larry Dignum

The Beaver, Autumn 1959

What are the chances of an enemy agent or a fugitive from justice hiding out in Canada's North? Pretty slim, for actually it's easier to hide in a crowd than in these great vacant spaces. Here the stranger may meet only prospectors, miners, or trappers, who may be Indian, Eskimo or white, but sooner or later word of his presence will reach the authorities through the Canadian Rangers, and his progress may be unaccountably delayed till he's been investigated.

The Canadian Rangers are an auxiliary force of the Reserve Militia under the Department of National Defence. They patrol the lonely places of Canada and are a potential guerilla army. They are an outgrowth of the Pacific Coast Militia Rangers, so a backward look at that little-known organization gives a realistic picture of this present one.

The PCMRs were organized early in 1942 as an auxiliary corps for defence against the Japanese. They were made up of residents in all parts of British Columbia and as far north as Dawson City in the Yukon. Among them were fishermen, trappers, ranchers, loggers, storekeepers, railway hands, miners and road maintenance men, white and Indian alike. All were trained to provide information, report suspicious activities and, if necessary, resist enemy landing, and their motto was Vigilance, Integrity and Silence.

Many of them were skilled woodsmen and hunters who could shoot, stalk, and track over any sort of country. The Canadian Government issued them with "dry-backs," loose canvas coat and trousers that shed the rain, and a wide brimmed fisherman's type hat. They were also provided with a rifle and an annual allotment of ammunition and every Ranger carried a rope fitted with toggles so it could be joined with other ropes.

The PCMRs were loosely organized in companies but carried on their regular occupations, functioning more as individuals than as teams. They trained on their own at week-ends, and periodically they were inspected by the general officer commanding at Vancouver. At first they attempted the conventional line formation with presentation of arms, but this was not very successful and a newly arrived [General Officer Commanding (G.O.C.)] on his first inspection of a company in the Fraser Valley suggested that they concentrate instead on scouting and stalking, practicing the techniques of surprise and cunning to confound an enemy.

Word passed quickly through the Rangers that the new general wanted scouts, not soldiers. The next time that G.O.C. inspected a company he arrived with his staff officer at the appointed time and place, a wide clearing in the bushland bordering the Fraser Canyon. To greet him was a lone figure in loose canvas clothing with rifle slung on his shoulder.

"Are you the Ranger Captain?" asked the general.

"Yes sir."

"Is this where I'm to inspect your company?"

"Yes sir."

The short stocky major-general surveyed the wilderness around him. There wasn't another human sound, nor soul in sight. He wasn't prepared for this and he fixed his piercing eyes on the man in front of him.

"And when am I to inspect your company, Captain?" he asked testily.

"Whenever you're ready sir."

The fiery little man exploded.

"I'm ready NOW!" he roared.

The Ranger captain raised his arm and a volley of rifle fire filled the air. Suddenly up from the ground, from behind trees, bushes and cover of any sort Rangers appeared and advancing in a body, formed a circle around the two staff officers and their captain.

The general was completely surprised and confounded but took the joke on himself with good grace and complimented him on their clever tactics.

The Rangers served without pay and there were no age restrictions, but most of them were either too old or too young to join the armed forces. On an all-day anti-sabotage exercise one Sunday the G.O.C. was invited to attend. He arrived with two members of his staff, and leaving the car on the road, started walking into the bush where the exercise was to be held. As expected, there was no one in sight, but as they rounded a curve in the trail a high cracked voice shouted: "Halt!"

From behind a large boulder stepped an ancient white bearded Ranger, rifle aimed steadily in their direction.

"Come for'ard one at a time and say who ye be," he ordered.

The general advanced and presented his credentials, then stood by while the other two officers did likewise. Having been cleared and recognized, the general started to chat with the old man who said he had been in the Rangers from the start.

"Wouldn't you have a better view of all approaches at a higher level?" asked the G.O.C.

"Aye, that I would," agreed the old fellow. "But me legs ain't good enough to climb. I gotta good eye, though, an' I kin shoot!"

"I'm sure you can," the general assured him. "And you're doing a fine job. Nevertheless I don't think you should be alone. Three desperate men could have overpowered you."

"Shucks! I ain't alone."

He blew a short blast on his whistle, and forty feet away from the little group a pint-sized Ranger rose up out of the ground, his rifle covering the three officers but a wide grin splitting his freckled face. Grand-dad and grandson, aged eighty-five and thirteen respectively, were both Rangers and represented the extremes of the age bracket.

The PCMRs served in a variety of ways. They patrolled regions where they lived and with which they were familiar, and any stranger in that area roused suspicion. The road workers, railway men, postmasters, Rangers all, noted the newcomer and passed the word along. Except in emergency, they took no action and travellers were rarely aware that they were observed all along the line.

The Rangers were specially valuable in search and rescue when aircraft were lost in the mountains, and also provided much needed topographical information, but they were never used for guard duty. This was the role of the Military, but they did take part in exercises with the Services, using their scouting skills and guerilla tactics to reveal weaknesses in defences thought to be secure.

In the spring of 1945 the Japanese launched incendiary balloons that drifted across the Pacific to this continent, but the close watch of the Rangers prevented any serious conflagrations in the forests, and they recovered so many bits and pieces of the balloons that the Royal Canadian Engineers at Chilliwack were able to reconstruct and study this gimcrack weapon.

When the Pacific War ended in 1945, the PCMRs were disbanded. But in 1947 National Defence Headquarters recognized the contribution they had made in guarding the West Coast, and decided that such a body of men would be of value in peace time too, and would provide a nucleus for rapid expansion if another war broke out.

Consequently the Ranger organization was revived but with a wider concept. Today from the Pacific to the Atlantic Oceans there are Rangers in isolated and sparsely populated areas where there is no other practical means of surveillance, and they do for all Canada what the PCMRs did for British Columbia.

Still functioning as individuals, they carry out their duties in conjunction with their regular jobs. The company unit is for control and administration, providing a two-way channel for information, often through Hudson's Bay posts. Even in peace time there is the need to report suspicious activities, to guide, to search, to rescue, and to supply vital information about the formidable country they live in. If necessary the Canadian Rangers will assist in local defence until the arrival of regular troops.

When on duty they wear a scarlet armband with the three maple leaves of the Canadian Army superimposed on a crossed rifle and axe. They have no uniforms, receive no pay, seek no glory, but these men of known loyalty, Indian, Eskimo and white, take pride in standing on guard in the empty and remote parts of Canada with vigilance and integrity, and in silence.

– 17 –

Outpost Defenders "Just a Bit Better Prepared": 2500 men of the Canadian Rangers are ready for almost any emergency

Canada Month, March 1964

In the North, down the bleak coast of British Columbia, and scattered in scores of out-of-the-way spots from one end of Canada to the other are little groups of civilians who, by natural bent or by particular experience, are a little better prepared to deal with emergencies than their fellows. They are the unsung men of the Canadian Rangers. If the country goes to war, they will report to local military commanders as scouts. If Canada is ever invaded without warning, they are ready to take to the hills as guerillas.

There are close to 2500 such men, whose second calling is hardly known to most of the people they encounter in daily life. They are organized in 42 groups ranging in size from company down to platoon. The larger units are split up in different locations, and altogether 148 towns and villages—the outposts of Canada, for the most part—have Ranger groups.

One such unit is in the Yukon's capital, Whitehorse. In an emergency its ranking officer, Kit Squirechuck, could summon 35 men—civil servants from Whitehorse offices, truck drivers from the Alaska Highway, construction workers, Indian trappers (a third of Squirechuck's group are Indians). Many of them are veterans, the rest are men whose life already equips them for unconventional fighting.

Squirechuck, a sergeant in the Rangers, is a civilian employee working in the Army's local ordnance shops as an armorer. A crack shot he, like others in the unit, has an army rifle and draws 100 rounds of ammunition a year from the Army. That's the full extent of the official physical support the Rangers get. No uniforms,—no pay; though some have had survival training for Arctic conditions with Regular Army units.

Heading the 14 Rangers in Dawson City is Don Shailer, who works in the local Department of Transport office. His men are much like Squirechuck's—trappers, woodcutters, truck drivers, construction workers, Dawson civilians who hunt for sport and who are familiar with bush living and firearms. Shailer expects that his group would be called on for scouting, and to advise on living and traveling in the near-Arctic cold of his area.

The Army makes use of its Rangers even without a war on. They are sometimes used in Army exercises in the role of the infiltrating enemy, and they take pride in their skill and their knowledge of local country, which often enables them to outsmart the ponderous Regulars. The men are officially part of the Army Reserve, who supply the rifles and the ammunition. Reserve also makes promotions, though Army headquarters in Ottawa admits readily that the units are fairly autonomous, put up their own commanders, and simply ask for official approval.

They are all unpaid volunteers, and most of their training is at the discretion of their commanders.

The Canadian Rangers were formed in 1947, when the Cold War was just beginning to make itself felt, and when the possibility of isolated actions in the North seemed more than just a far-off possibility. And no one has seen any reason to stand them down. As Ottawa says: 'We can have Ranger groups in many places where it would be much less convenient and economical to have Regular Army units. And they're just as much a requirement today as they ever were.' The Army points out, too, that the men offer a great deal of assistance to the RCMP and other police in emergencies.

Actually, there is good reason to encourage the Rangers right now, when the Department of Defence is pulling back from the Arctic and sub-Arctic for the sake of economy.

Not everyone at Defence is happy about this. The Defence Research Board's R.J. Sutherland points out that Canada's legal claim to its Arctic islands is not established without question, and that recognizing a responsibility for defence there is one way to confirm it.

That is probably not something the Rangers can do. But as the Army pulls out, some of the Rangers are hopeful that some funds might be thrown their way for organization and training. Whitehorse, for instance, will probably feel keener about the Rangers when the Army is gone.

CANADA MONTH correspondents sent back a lively picture of the Rangers. Jack Worsell made an observation that might stand for all of them: 'Banish, with regret, the picture of a lone figure on snowshoes, rifle in hand, patrolling the frozen wastes, a shadowy, heroic figure in the dim light of the sub-Arctic sun. These are just normal types, from every walk of life, with just a slight edge on their fellows in the matter of preparedness. The slight edge that might make the difference between utter defeat and a fighting chance.'

– 18 –

The Shadowy Force on Guard in the Arctic

Scott Young

Globe and Mail, 14 April 1969

The scene is a Hudson's Bay Co. post in the high Arctic. A man lets his dogteam rest a few yards away and walks in, pulling a notebook from his parka. He tells the Hudson's Bay man of something he saw on this trip just completed—a flash of light in the sky, wreckage on the beach, or maybe a polar bear with the initials CCCP (which is Russian for USSR) embroidered into its rear end. The trader goes to his two-way radio and relays the information. An hour or two later it lands on the desk of a lieutenant-colonel in the Defense Department in Ottawa…

Diefenbaker suggestion

Does this sound like an extension of former Prime Minister John Diefenbaker's suggestion last Thursday that Canada should set up a force of Eskimos to help guard our Arctic?

It does. Perhaps more pertinent is that we have had a force precisely of this nature for nearly 22 years. Right now an interdepartmental study group involving Defense and other departments is reviewing Canada's military presence in the North. The role and organization of the 1,683-man Canadian Rangers, volunteer Eskimos, Indians, Metis and Whites, are part of this study.

By coincidence, at almost the exact time that Mr. Diefenbaker in Whitehorse was visualizing his Eskimo force, I was in the Defense Department trying to get information about the Rangers. One thing you have to give them credit for is stealth. Not many people know about them. Even Mr. Diefenbaker, one of the most North-conscious of all Prime Ministers, either wasn't aware of them or his references to them were misunderstood as being suggestions for a new force.

Need little training

I'd heard about the Rangers the day before while seeking information on exactly how many servicemen are stationed in the North. The figure is about 500, mainly in signals. "But then there's this other outfit, the Rangers," said a colonel who spends his days mainly considering the most advanced electronic defense devices of the age. "They don't get any training—but then they're born with most of the training they need. I think we give them each a rifle and a few rounds of ammunition, but that is about all I know about them."

The following day, with help, the lieutenant-colonel was located who had all the answers about the Rangers. He did not wish to give them by telephone, however. He wanted my questions to be typed out and sent to him, whereupon he would have the answers typed out and sent back. This was done. From here on everything is very official.

The Canadian Rangers were set up by order-in-council on Sept. 4. 1947. Brooke Claxton was Defense Minister at the time, one when deep distrust of Soviet intentions had replaced the friendship and co-operation of wartime. For this reason there was felt to be a need for some kind of an observer corps in the Arctic, organized on army lines. The first company established was based in Dawson City and the second in Whitehorse; right where Mr. Diefenbaker was speaking last Thursday.

The 1,683 Rangers currently on strength are almost all Eskimos, Metis and Indians, but also include a few employees of Government departments and of civilian industries. All are volunteers, both officers and men, committed to be ready for action in time of emergency. There is no pay—although they may be paid if they are asked to perform specific tasks or undergo specialist training. Each Ranger on joining receives one .303 rifle, 300 rounds of ammunition, an armband and a notebook. Annually thereafter he receives 100 rounds of replacement ammunition. There have been no strikes for higher pay.

The role of the Rangers obviously has been thought out with care. They are organized in 28 companies, each with its own headquarters, and each with varying numbers of platoons. A company area may be hundreds of square miles of sparsely settled and isolated north that couldn't conveniently or economically be covered by other elements of the armed forces. In a few cases Ranger units have been assigned definite local defense tasks, but the readiness and observation aspect is more the normal line of duty.

Use any system

In reporting to Ottawa or communicating with other Ranger units, they use whatever signal system is handy—Government or civilian communications such as Department of Transport, Royal Canadian Mounted Police, and the Hudson's Bay Co.

In addition to reporting suspicious activities that may occur, the Rangers provide guides for regular force units that may come into the area. They also perform spotting duties, including locating and reporting aircraft in distress: collect detailed topographical and other information for use by the regular forces: perform coast-watching assignments: and man rescue parties when required. Their work extends from one company on the east coast of Newfoundland (which is about as far east as Canada goes) clear across the North. There are two companies in Labrador, four on Baffin Island. Headquarters names all have the ring of the North—among them Sugluk, Rupert House, Pond Inlet, Frobisher Bay, Port Radium, Destruction Bay, Alexis Creek.

General statement

Ten days ago Prime Minister Pierre Trudeau announced Phase One of the revised Canadian defense policy: a general statement of aims.

Some time in the future he will announce Phase Two: the detailed plans.

Recommendations have been made to the Cabinet that more work, training, pay and attention be given to the Canadian Rangers, and it is possible that a tiny item in Mr. Trudeau's Phase Two will announce expansion of this almost-unknown force.

Perhaps the essential Canadian-ness of the Rangers' role will help them compete in Cabinet with all the mighty hardware of other proposals.

– 19 –

The Canadian Rangers: Men on the Spot

Sentinel 6/10 (November-December 1970)

Last summer, a staff officer from Mobile Command flew some 12,000 miles in *Otters* of the Air Reserve, in a zig zag course which took him from the outposts of Newfoundland through a wide loop into the Arctic to the narrow fiords of Ocean Falls on the west coast. Enroute, he visited 27 communities which had units of the Canadian Rangers.

Major Bill Stirling was charged with reporting on the current status of the Rangers, assessing their usefulness and making recommendations of their future. It was a long trip.

The Canadian Rangers were a direct successor to the Pacific Coast Militia Rangers of the Second World War. Back in 1942, when Japanese power in the Pacific posed a very real threat to Canada's west coast, the PCMR were formed as a body of coast watchers, aircraft spotters, and local defence forces along the B.C. coast and up to the Yukon. They performed a very useful function.

Then, in the immediate post war period, the concept was expanded to include all of the remote areas of Canada, including the Canadian Arctic, and the Canadian Rangers were born. The point was that they were the men on the spot – Indians, Eskimos, Métis, and white men – who were ranging the areas around the outposts of civilization or the vast areas of Canada which civilization had not yet touched. They knew how to live, and move and work in their areas. They would be useful to the military.

They were given such tasks as guiding regular troops, providing a reconnaissance screen, and coast watching. They were to help the RCMP and Provincial Police (where applicable) in apprehending small groups of enemy agents. They were to report on friendly aircraft in distress and help the crews of such aircraft, and they were to report on enemy aircraft. Finally, if the situation required it, they were visualized as the nucleus for guerilla bands operating against an enemy invader.

As the years went by, the Rangers were not forgotten, but the need for them became progressively less. New and sophisticated means of detecting enemy aircraft made their role as aircraft spotters almost redundant. New and improved communications into the north and on the two coasts eroded their usefulness as coast watchers. Long range aircraft and increased military and Government activity in the remote areas of Canada filled the surveillance role better than the Rangers could. An improved and expanded search and rescue organization took most of the load off the Rangers for reporting and helping downed aircraft. In short, as civilization rolled back the frontiers of Canada, the need for the Rangers decreased.

At the same time, the day-to-day lives of the individual members of the Rangers were changing. The Indians and Eskimos concentrated more and more in communities, and fewer of them were hunting, fishing and trapping. They covered less and less territory and for many of them their old skills of living and moving in severe temperatures became rusty. And during the same period, the Army, Navy and the Air Force were becoming more and more familiar with Arctic conditions in their operations, exercises and explorations.

So, by the time Major Stirling made his grand tour to report to the Commander of Mobile Command, many of the units which had once constituted the 1600-man force were practically non-existent, while others were still relatively active. As recently as 1969, Mr. E.J. Galibois, commander of 15 Company, Canadian Rangers, produced 30 Rangers to act as "enemy" and guides during Operation Top Hat.

But Major Stirling had a good look. Flying in a 401 Reserve Squadron *Otter* out of Montreal, he started his investigation with Anticosti Island, and 3,500 miles later he had covered the outposts in Newfoundland and Labrador. An *Otter* from 400 Reserve Squadron in Toronto took him up to Moosonee and Moose Factory on James Bay. Another *Otter* from 402 Reserve Squadron in Winnipeg flew him some 5,000 miles up the Northwest Highway System, through the Western Sub-Arctic and down the Mackenzie River to the Arctic Ocean. And finally, a 418 Reserve Squadron aircraft from Edmonton flew him into the more remote communities on the west coast.

Wherever Major Stirling went he got the same impression; that the roles of the Rangers needed revising.

Then, late in the summer, both Major Stirling and the responsibility for the Rangers passed to the new Northern Region Headquarters. And this is where the future of the Rangers lies at this moment.

But whatever the future holds for the Rangers, Brigadier-General Ramsey Withers, Commander of Northern Region Headquarters, is determined that the force should reinforce the pattern of development in the north. He wants to talk to a great many more people in the Arctic and sub-Arctic before he comes to any conclusions about this proud force of unpaid volunteers.

– 20 –

Our Northern Eyes...

Sentinel, November-December 1971

The three-ship task group which deployed to the eastern sub arctic last summer meant a new mission for the Maritime Command.

While the ships were landing supplies at remote settlements in Hudson Strait, Ungava Bay and Labrador, their helicopters were flying three army officers to outposts of the Canadian Rangers.

As well as bringing the Rangers their own supplies, the three discussed with them matters of communications, arranged further supply visits and a "host of other details," said the Ranger liaison officer embarked in HMCS *Preserver*, Captain Bruce Fraser. The three liaison officers are stationed normally at CFS St. John's, Newfoundland.

Rangers are a sub component of the militia. They originated in 1942 when Japanese power in the Pacific posed a very real threat to Canada's west coast. They were formed as an organization of coast watchers, aircraft spotters and for local defence along the B.C. shoreline. The Pacific Coast Militia Rangers were unpaid volunteers making a useful contribution to the country's security in the war. In later years the organization was somewhat neglected.

With the recent birth of the northern regional headquarters of the Canadian forces at Yellowknife, it took control over northern central Rangers. Maritime Command became responsible for Rangers east and west.

Today Rangers still provide a small military presence in sparsely settled northern coastal areas of comparative isolation in Canada. They carry out coastal surveillance and report any unusual activity to their liaison officers in St. John's. They report on and aid vessels in distress, do the same in the case of aircraft and provide expertise on local terrain in event of forest fires and other incidents.

Rangers are respected members of their communities. All have fulltime work, chiefly as hunters, fishermen and trappers.

Major Bob Bridgeman in the *Assiniboine* had done Ranger resupplying before. So he was made responsible for much of the planning and control of Rangers assigned to Maritime Command. He expects to visit Rangers in northern Quebec and Labrador at least twice yearly and those in Newfoundland more often than that. His territory outside Newfoundland goes from Baie Comeau up the Labrador coast to Ungava Bay, down the east coast of Hudson Bay to Povungnituk. Looking back on last summer's outpost tours, the major forecast: "in a couple of years the Ranger organization will prove to be a viable force in the north."

– 21 –

Bannock Binds a Better Team

Capt Craig Mills

Sentinel, February 1975

Bannock!

The ancient Scots came up with a flat disc of unleavened bread which was noted for nutrition, flavour and longevity. Brought to North America by early Scottish explorers, it was adopted and, some claim, improved upon by native Canadians.

It's the best bread made. Just ask Olayuk Naketaquik, or Charlie Jim Nitsiza. They're Canadian Rangers, residents of the Northwest Territories, and they should know.

For that matter, ask Master Corporals Joe Martin or Donald Gariepy. They are members of 3rd Battalion, Princess Patricia's Canadian Light Infantry, and are residents of Victoria — and they should know too.

In fact bannock, that delightful northern concoction, may have helped cement relations between the Forces and the Rangers.

Different bands meet

It all began last March, when Northern Region Headquarters in Yellowknife started its most ambitious Rangers training exercise to date. It was named Nanook Ranger II.

Fifteen northerners gathered at Camp Antler, a converted summer resort camp, 40 miles west of Yellowknife, to meet an officer, a sergeant and six keen, young master corporals from 3 PPCLI—their instructors for the next month.

Five Eskimos from Arctic Bay, 1,200 miles north-east, and six from Holman Island, 540 miles north-west flew in for the training. Four Dog Rib Indians spent five days on the trail travelling by dog teams from their homes in Lac La Martre, 75 miles away. This marked the first time for Indian and Eskimo Rangers to train together.

Enough problems

Any pessimist could foresee the problems. Only one Dog Rib could speak English. None of the instructors had ever worked with Rangers. The Eskimos rarely worked or lived this far south of the tree line. At first, no one liked the RP4 and freeze-dried rations. The dogs needed plenty of fish and meal. The snowmobiles were finicky, and shacks were too hot and confining for the Indian Rangers.

Age and experience differed vastly. John Avilingak from Holman Island, 16 was the youngest Ranger, and this was his first exercise. Issac Attagutsiak, from Arctic Bay at age 56, was the oldest and most experienced Ranger with 25 years continuous service.

How could it possibly work? In one month, Lieutenant Mike Lawless, course training officer, Sergeant Jim Clare, his second-in-command, and six instructors were supposed to turn out trained patrol leaders. Course subjects included weapon handling, map using, patrolling, first aid, radio communication, and many other necessary skills. Was it possible to teach these subjects, conduct a week-long patrol, and solve the myriad administrative problems all in a month? A small miracle might help—and it did.

Bannock banter

Bannock just may have been the thread that tied it all together. Early in the training, one Ranger, tired of chewing the spongy glop most Canadians call bread, took matters literally into his own hands. Using gestures sparingly, and no words at all, he demonstrated the ancient art of making bread, Northern style, to course cook, Master Corporal Howard Dowswell.

It would be understating somewhat to say that the cook's first efforts did not quite reach epicurean heights — but time, determination and practice finally brought forth a product resembling bannock.

It proved an instant success and tremendous morale booster. A common ground had been found — everyone liked it. Military rations became more palatable with it, game and fish tasted better than ever, and it proved especially delightful for tea dunking.

Needless to say, bannock was not the only tool of success. Instead of a strict instructor/student relationship, the NCOs and Rangers began to share all of their knowledge and skills.

Who are the teachers?

Instructors lived with their own patrols, each from a different settlement. Rangers learned and improved their English, instructors learned a smattering of Dog Rib and Eskimo dialects — even though many of the words and phrases were not usable in mixed company.

Rangers mastered the vagaries of the snowmobile — and one Holman Island Ranger, known as "super-wrench," could strip, repair and assemble an oversnow vehicle better than anyone.

Instructors learned trapping and fishing techniques, mainly from the Lac La Martre Indians. Master Corporal Ted Luscombe became especially good with the dog teams, once giving Brigadier-General [J.A.] Fulton, Commander, Northern Region, a ride he'll not soon forget.

All Rangers took to skiing and snowshoe broomball like they were born to the sports. They proved to be expert riflemen and naturally, masters of survival in the bush.

Lieutenant Lawless put it simply "We teach each other—and we understand each other."

Ranger background

So what does it all prove?

When the Canadian Rangers were established in 1947 as a sub-component of the Reserves, it was planned to use native northerners, mainly hunters and trappers, for reconnaissance, surveillance and guiding. Rangers were to work individually, without pay, rank or title, reporting unusual incidents to authorities in the course of their normal travels. They received a rifle, and ammunition each year. However, they attended no annual training, and once properly enlisted were virtually on their own.

Periodic visits from military authorities became scarce after several years and the Ranger program appeared to languish until April, 1971, when Canadian Forces Northern Region took over responsibility for Ranger training.

The basic concept remained unchanged, but an active program to reconstitute the Rangers got underway immediately.

Major Bob Lemaire, staff officer in charge of Ranger training at Northern Region, has been at it now for three years and he is "still learning from them—every day."

Major Lemaire and his staff started sifting through over 400 documents of former Rangers. Visits to settlements were begun to find active Rangers and willing recruits. Training programs began at Holman Island, Lake Harbour, Lac La Martre and Frobisher Bay.

Three exercises were conducted for Rangers, and several Rangers took part in Exercise Patrouille Nocturne, near Frobisher Bay in February, 1972.

Finally came Nanook Ranger II. Each Ranger was given acting sergeant rank, because this was a patrol leader's course (except for Isaac who holds substantive sergeant's rank). Rangers are paid according to their rank while training with the Regular Forces. In addition, each Ranger is paid rent during exercises for the use of his snowmobile, sled and fuel, or dog teams and food.

This mixing of Eskimo and Indian Rangers on the same exercise was an experiment.

And it worked.

Cultures can mix

Oh, sure there were areas to sort out, on equipment scales, rations, operation methods and teaching skills. But it did work!

Indian and Eskimo Rangers proved they are able to work with each other and with a military force successfully.

Trained, competent Forces personnel are learning how to employ the Rangers' unsurpassed bush skills and how to live harmoniously with Northerners.

Nanook Ranger II proved that different Canadian life styles can be compatible and can help to build a better country.

Isn't that a worthwhile objective?

– 22 –

Canadian Rangers Stand on Guard

David Miller

News of the North, 7 November 1979

What if a grim historical pattern repeats itself, global conflict breaks out and the North is threatened by over-the-pole invasion?

Who would stand on guard for us until the arrival of reinforcements from the South?

The Canadian Rangers, that's who -- a rough-and-ready bunch of unsung latter-day heroes, our shadow army of the North and, according to DND military strategists, probably the most cost-effective armed force in the Canadian military.

In 23 home communities scattered thinly across tundra and taiga, a total of 450 Ranger reserves stand as our northern sentinels, maintaining a vigilant eye over our nation's Arctic approaches.

In the event of hostile foreign intrusion, these locals would emerge from hamlet and settlement to confront the enemy guerilla-style until a Canadian vanguard of regulars could mobilize and deploy. Following that, the Rangers would serve as local scouts and back-woods warriors.

This has been the basic role of the Canadian Rangers since their inception in 1942. Numbering over 18,000 at the height of the Second World War, they were our response to a Japanese Pacific coast invasion threat, our own wood-hewn version of the British Home Guard.

With peacetime security, these Pacific Coast Rangers dissolved. But as the Cold War hardened, the Canadian Ranger idea underwent a northern resurrection with the organization of squads at Dawson City and Whitehorse. Two years later, Yellowknife became the headquarters of the 7th Ranger company, one of three designated for north of 60.

The ranks of the Canadian Arctic Rangers reached a maximum of 1,600, before waning in the late fifties as the emphasis shifted away from conventional ground defence. In the lull of Detente, their numbers declined ever further.

But while the history of this frontier force ebbed and flowed roughly with our national paranoia, the quality of recruit has not changed. They are selected for their unique wilderness survival skills and their detailed knowledge of the local territory. Parade square potential and spit and polish appearance do not mark a man for service in this special Arctic guard.

Most Canadian Rangers are rugged hunters and trappers. Some nine out of ten are Inuit, Dene, or Metis. All are humble servants of the Queen.

But in spite of patriotic loyalty, "you don't order the Rangers out: instead you ask them -- and kindly," explains Northern Command operations officer Sandy McDonald.

And then, you still don't know how many will show up. If it's in the middle of hunting season, they may just say 'bugger-off'. On the other hand, if it's a slow time of year you may have to beat 'em off with a stick."

Except under national emergency, only the Governor-General has authority to order them up.

"But where could you get troops of that potential -- very handy militarily -- for the cost of an old .303 rifle and 300 rounds of ammo a year," McDonald remarks.

That's all the pay they get for their vigilance. (They draw militia wages when on manoeuvers.)

The Rangers have no uniform, except for an arm band -- the minimum required by the Geneva Convention to designate them as combatants as opposed to terrorists, spies, or partisan fighters.

This band is scarlet with a green patch bearing crossed rifle and woodsman's axe on three maple leaves.

When Arctic Rangers go on parade, the custom of wearing Eskimo parkas—hoods up – has been adopted. This tradition originates from Frobisher Bay in 1956 when they formed an honor guard for former Defence Minister C.D. Howe and his U.S. counterpart on joint Arctic defense inspection tours.

In our own north of 60, the Canadian Arctic Rangers are organized on paper into battalions, companies and patrols. But on the ground, only the patrol unit carries any relevance. At this level, in true Spartan style, squad commanders are appointed according to proxy of rank and file. Inevitably, patrol chain of command parallels the community's pecking order, says Major John Tattersall, their senior commander in the North.

He likens them to the original pre-Confederation Canadian militia. But instead of discarding plow for musket to fight a Fenian foe, for today's Rangers, its trap line for land mine.

His colleague, Capt. McDonald, calls them "a bit of an anachronism".

"The cynicism of today isn't there. They believe they're really defending their homeland. They're sincerely loyal, like to be associated with the armed forces and are very proud of it."

While the Rangers may seem a bit out of date, their usefulness is more relevant than ever.

Aside from their defence mission in event of war, they perform a number of valuable duties. As guides intimately familiar with the local environment, they have been employed to assist in every major Canadian military training operation in the North.

Ranger training is usually updated once a year.

"However," explains Capt. McDonald, "it's pretty hard to instruct them in the field. They're usually instructing the instructors.

"The air force Arctic survival school uses Rangers to show how to skin a seal, snare a hare, build an igloo."

The Rangers' special skills are also of value in RCMP and DND search and rescue exercises.

Along with watch-dog missions performed in the course of their everyday lives, the Canadian Rangers bolster Canada's tenuous Arctic sovereignty. In areas which could not be covered conveniently or economically by our regular armed forces, they provide a military presence.

But the effectiveness of the Rangers is endangered, explains Capt. McDonald.

"As the old ways and the traditional skills die away, places are forced to get out of the Ranger business. You're more likely to find Ranger material in places like Holman Island than in Yellowknife...People are losing those special talents which make the Rangers so valuable."

"In response, we may have to go over to a more formal organization as the old skills drift away."

McDonald says the Department of National Defence thinks the Rangers could form the nucleus of a more regular, permanent, northern-based reserve or militia.

But he advises caution.

"In Alaska they had something similar to our Rangers -- the Alaska Scouts. When they decided to turn them into a formal national guard they destroyed them because all they had time for then was polishing boots and sending memos.

"There's a danger that over-organization could ruin the effectiveness of our Rangers," he says.

The Northern Command may soon have to juggle this dilemma of organization versus extinction.

Northern commander General Clay Beattie recently announced that a comprehensive report on Ranger potential is presently underway with a view to reinvigorating this unique armed force.

He has been discussing an expanded Ranger role with his colleagues at defence headquarters in Ottawa.

"I see an important role for the Rangers in the future of the North...new opportunities for them to co-operate in mobile command training exercises... increased numbers. In general, we want to raise the level of their capability," Beattie said.

This isn't the first time the Ranger idea has been pulled down from the attic, though. If concrete moves aren't made soon to save them from decline and reinforce their special skills, only the last catch word in their motto -- Vigilance, Integrity and Silence -- may carry any meaning.

– 23 –

The Raw Reality of Staying Alive

Ken Spotswood

Edmonton Journal, 22 September 1981

Resolute Bay, N.W.T.—It was not a pretty sight for the hardy Canadian Forces soldiers as the dark blood trickled through the ice of the frozen seashore.

They stood and watched in silence as the victim—a seal—was laid open under the expert knives of two Inuit members of the Canadian Rangers.

Cheating death is the key to life here in the North. Keeping body and soul together depends on keeping a cool head and learning to use every available resource.

For the 30 members of Charlie Company, 2nd Battalion, Princess Patricia's Canadian Light Infantry, this was a four-day lesson in survival that none would forget.

Stomachs growled and some choked back the dry heaves as the seal was systematically turned inside out. The real shocker came when the Rangers ate the raw meat.

There's no wood for fuel in this frozen desert. Raw meat is often the only alternative to starvation.

At first reluctant, the soldiers soon took turns sampling the flesh, remarking on its delicious flavour—particularly the liver.

Nothing is wasted. The meat was portioned out for a later meal and the thick layer of blubber was rendered into oil for heat lamps and cooking. The lower intestine can be used for fish and animal bait.

Sleeping in 10-man army tents is a luxury a stranded person would not normally have, so the course turned to shelter.

Again the Rangers displayed their superior knowledge of the land by building a two-man igloo out of a snowpack that at first looked too small to provide adequate housing.

The Rangers are largely Inuit and Indian hunters or trappers and number between 500 and 600 throughout the North. They are recruited for their knowledge of the areas in which they live, and for their special skills as guides and interpreters in Arctic operations.

The two accompanying Charlie Company were 56-year-old Levi Nungaq, a hunter and carver, and Ludy Pudluk, who also happens to be the local member of the territorial legislature for High Arctic.

Both men live in Resolute Bay and were the centre of attention with their amazing Arctic skills.

They spotted ptarmigan where there appeared to be none.

They pulled Arctic char out of 12-Mile Lake with ease.

And to prove their mettle, they even ate army rations.

The survival course was part of Operation Sovereign Viking, one of the largest and most expensive military exercises ever deployed in the Far North.

For two months, close to 400 members of the Patricias were exposed to the sub-zero temperatures and isolation of the Canadian Arctic for a variety of purposes.

They conducted a series of high Arctic reconnaissance patrols on Cornwallis Island using M-113 armored personnel carriers and CH-135 Twin Huey helicopters.

They ran extensive and successful communications tests in a battle with the magnetic north pole, which normally sucks radio waves into the ground.

They tested a variety of weapons and modified equipment while providing support for a team of Canadian biologists.

Last, but not least, they also established a firm military presence of Canadian sovereignty in the Arctic islands.

"We now have a wealth of information on the ability of our vehicles to traverse all types of terrain, as well as tremendous confidence in our communications equipment, and our soldiers' ability to adapt to such a severe environment," said exercise commander Major Pierre LaChapelle.

Recent media reports that Soviet nuclear submarines may be casually cruising under the Canadian Arctic undetected do not bother Major LaChapelle.

He did acknowledge that if there was a nuclear strike from Canadian Arctic waters, those missiles would probably go right over Canada to U.S. targets.

"And if a 1,000 man enemy force was to land in the High Arctic, all we'd have to do is spray them with water and they'd freeze to death. It's as simple as that," he said.

– 24 –

Rangers of the North

Capt Jean Marcotte

Sentinel 1982/6

Outside the arctic tent, the mercury falls below -40°C. Inside, the sleepers' breath congeals to form layers of frost on the outside of their sleeping bags.

After waking and dressing in the frigid cold, Major John Tattersall picks up his parka and steps outside.

The air is dry and pure. The white contours of the distant mountains slowly brighten in a halo of light filtering through the early morning mist. But Tattersall is not fooled by appearances. He knows how deadly nature can be in this region for those unfamiliar with it. Three years have taught him that no one knows nature better than the men he is with today: Indians and Inuit of the Canadian Rangers.

As Northern Region's staff officer in charge of the Rangers, Tattersall has been organizing and directing patrols in the far north for three years. "I have learned to appreciate the value, knowledge and resourcefulness of our northern peoples. The Forces must learn from them to ensure our ability to live and fight in the north," he says.

Little Known History

The term 'Ranger' goes back to the Second World War, when the Japanese threat hung over the west coast of Canada. The government decided to create a militia to provide additional protection in the area.

Volunteers simply had to live on the British Columbia coast and agree to defend it. Each man was given a rifle and 100 rounds of ammunition. Their mission was to watch the sky and sea, and defend their compatriots if necessary.

Officially named the Pacific Coast Militia Rangers, by war's end ... members patroll[ed] the B.C. coast as far north as the Yukon. With the end of the war, the program was forgotten. But since the early 1970's, the Rangers have been revitalized.

Today, in almost 60 villages from Newfoundland to the Yukon, more than 1,400 people proudly wear the armband of the Canadian Rangers. Like their predecessors, they are all volunteers and local residents. Each man is given a .303 rifle and 200 rounds of ammunition a year.

Their mission is to keep their eyes open and report any unusual activities in their territory. They must be thoroughly familiar with their community and region so they can assist the Forces when necessary. And the Forces pays them whenever their knowledge and skills are used.

Testimony to their effectiveness is not hard to find. Veteran paratrooper WO Ralph Goebel, of the Special Service Force's reconnaissance platoon, worked with three Inuit Rangers last year. He is generous with his praise.

"Without the Rangers," he says, "my mission would probably have failed. The native men know their region. And, their equipment and work methods are in tune with the realities of the north. I learned a great deal from them."

Experience and then Some

To take advantage of this expertise, Maj Tattersall and his second-in-command, MWO Bob McMinn, try to organize Ranger exercises in two different locations each year. This year, a patrol covering 100 km was organized to train the villages of Fort McPherson and Aklavik at the same time.

In the Camp

Tattersall was accompanying that patrol. The camp, sound asleep only a moment ago, comes to life around him. Snowmobiles putter, men call to each other, and through the rough canvas of the tents, dots of red appear where the stoves are glowing.

The chaos is brief. Climbing onto their modern-day mounts, the Indians have already left to go hunting. But David Thompson remains behind. His snowmobile travelled only a few metres before it stopped dead.

Without a word, he firmly plants his feet on the ground and heaves his snowmobile onto its side. The transmission is broken. A hundred kilometers from his village, and thousands more from the nearest dealer, he has little choice: repair the vehicle or abandon it.

Looking unconcerned, Thompson enters the tent of Herbert Blake, the oldest member of the group and the leader of this patrol. He warms himself up before beginning repairs. Around the stove they talk about the camp, the day's hunting and David's problem. Moments later he re-emerges to face the icy cold.

In -40°C weather David removes the transmission. With his bare hands he dismantles the transmission piece by piece. The ball bearings have given out. He works alone, with great care and patience, oblivious to the temperature. For almost seven hours he slaves, transferring bearings from one of the rear wheels to the transmission.

WO Jim Peterson, a Med A at NRHQ and a volunteer on the exercise, was astounded. "I don't know any military mechanic, no matter how tough and competent, who could have done the same job." But David Thompson's expertise is not unique. Many native people are expert mechanics. They can repair almost anything, using methods others would never imagine.

While Thompson repaired his snowmobile, MWO McMinn, Capt Don Pippolo and seven Inuit from the village of Aklavik entered the camp. Capt Pippolo is responsible for the cadet program in the North. "The Rangers are worth their weight in gold," he says. "If you show them a point on a map and describe the countryside a little, they will guide you there in no time flat."

White Man's Time

As much as possible, the regular soldiers try to familiarize the Rangers with CF equipment. So, during the five days before the patrol set off, Tattersall and Peterson explained the use of maps and radio equipment, familiarized the men with military tents, stoves, weapons and rations and taught the basics of first aid.

But the concepts of discipline and time are less important to some northerners and Tattersall and Peterson found this particularly frustrating work.

WO Peterson remembered the rifle competition that he and Maj Tattersall arranged. "Everyone agreed to meet at 8:00 a.m. The next morning the Major and I arrived early and waited. After an hour or so, small groups appeared and wandered off again. At times, almost everyone was there," Peterson recalled. "Then a few minutes later there would be only a couple. Finally at 10:00 a.m., with some still missing, we began the competition. It was always like that."

But MWO McMinn, who has been working with the Rangers four years, adds that once they leave on expedition, fantastic changes take place. "The Rangers are trustworthy and full of energy on the trail. Once enroute, they look after everything with the utmost of care and an unparalleled sense of duty." McMinn says that seeing the Rangers work in their natural element will remain one of the greatest memories of his career.

While on these exercises, Maj Tattersall and MWO McMinn evaluate the Rangers' abilities. It is important for them to know the men well, since they must select the right guides for units which come to train in the north. Some are excellent mechanics or hunters, while others know a specific region particularly well.

MWO McMinn says that with time, the native people have learned to respect him and the others for what they are. "After all," he says, "few organizations send people to live up here to share and learn northern lifestyles. There are fewer still who absorb that knowledge in order to put it into practice for themselves."

He believes that over the years, Canadian military men have created a hardened reputation throughout the northern villages.

Like others, McMinn hopes southern-based units will make greater use of the Rangers. "The Rangers have an extra-ordinarily practical knowledge of an unfamiliar and unforgiving environment, and that knowledge should be put to use."

– 25 –

East Coast Rangers

Capt Richard Moore

The Sentinel 1984/3

"We don't issue rifles so they can roam the countryside, rifle on shoulder, defending their communities like some irregular militia," Captain Les Palhazi states emphatically. "The rifle is issued for target practise so they can participate in rifle shoots. Rifle shooting promotes esprit de corps and is a visible sign that they belong to something."

That 'something' is the Canadian Rangers. Capt. Palhazi, an infantry officer who previously served with the Third Battalion, Princess Patricia's Canadian Light Infantry, has been the Canadian Ranger liaison officer in St Johns Nfld., for the past three years.

He didn't know much about the Canadian Rangers before taking up the appointment. "There's no course to prepare you for this job," Palhazi says.

But, he adds, it didn't take him long to learn. Shortly after his arrival in St John's he boarded the naval reserve vessel HMCS *Fort Steele* and sailed in her all the way to Nain, Labrador, visiting Canadian Ranger units along the way in coastal communities on the Newfoundland and Labrador coasts and Quebec's lower north shore.

During the visit, Palhazi learned that there are more than 700 east coast Rangers. They come from every walk of life but the majority are outdoorsmen – fishermen, woodsmen, fishplant workers, bushpilots, surveyors and fisheries officers. Unlike northern region's Rangers, only about 100 of the east coast Rangers are native Inuit or Indians.

Service is voluntary and unpaid unless Rangers are on callout or attending a training seminar. Their average age is in the 40s but they can stay up to 65. Palhazi says there really isn't any set compulsory retirement age but gentle persuasion is used to persuade members to retire at 65 in order to renew the program.

Even though they are a sub-component of the CF reserve force, aside from an armband, a Ranger ID card, a .303 rifle and 200 rounds of ammunition yearly, the Rangers wear no uniforms and cannot be distinguished from the civilian population.

As Capt Palhazi says, Rangers are not defenders of their communities. They are an informal organization of coast watchers and an initial point of military contact in remote and isolated coastal communities along Canada's eastern seaboard. They report on local conditions and unusual activities along Canada's coasts, report search and rescue incidents and act as guides for Canada's regular and reserve forces.

Capt Palhazi says that there's a lot of travel involved in administering the program. To reach his far-flung charges he uses a wide variety of transport. Everything from chartered aircraft, CN ferries or his four-wheel-drive military vehicle are used to get him to the troops.

As he travels, Palhazi meets new and old Rangers, distributes rifles and ammunition, supervises rifle shoots, delivers ID cards and armbands and presents awards and retirement certificates to deserving members. He also attends many of the Ranger social functions because his visits often turn into a social occasion. When there isn't a local motel or hotel available he stays in the house of a friendly Ranger.

Palhazi is especially proud of two programs he has initiated. It didn't take long to realize that one of the biggest complaints from Rangers was a lack of recognition for their efforts. After investigating, the young officer realized the Rangers had a point. As a result, in May 1982, he received permission to begin awarding Canadian Forces decorations to qualified Rangers. Since then almost 400 have been presented, many with second clasps. These presentations have become important events in the Ranger community and in many cases result in front page coverage in the local press. The same goes for presentation of certificates of service to retiring Rangers.

Capt Palhazi is also pleased with the four regional training seminars he holds yearly for Rangers.

The job of Ranger liaison officer is as unique as the Canadian Rangers themselves. Palhazi is certain "there's no job like it in the Forces." After three years of it he should know. So if you like the outdoors, travel, lots of work, Newfoundland and meeting people this could be a job for you. All you have to do is think Canadian Ranger.

– 26 –

Eskimo Point Rangers Honoured For Bravery

Brian Mitchell

news/north, 20 April 1984

Eskimo Point—Since the Canadian Armed Forces conducted manoeuvers here a year ago, the Inuit have adopted a strong military presence in the community.

Until three months ago, 16 Canadian Rangers armed with Forces' issue rifles were the military eyes in this part of the Keewatin. Now there is a full complement of cadets.

Two of the Rangers, Peter and Johnny Mamgark, have been awarded one of the highest military commendations for their actions in saving the lives of a patrol during Exercise Kovick last year.

The presentations of the Chief of Defence Staff commendation were made last week at the community hall by Northern Region Headquarters Commander, Brigadier-General Mark Dodd.

The community hall was packed to overflowing for the formal ceremony involving 11 Rangers, 36 cadets and their leaders.

Standing at attention, nerves fluttering, the row of Rangers flanked by three rows of uniformed cadets were inspected by General Dodd. Later, the cadets and Rangers were praised by the General for their deportment and poise. It was also noted that the group had to be re-supplied with black polish.

The prestigious commendation awarded to the Mamgarks may be awarded to any member of the Canadian Armed Forces who has performed a deed or activity that is considered beyond the demands of normal duty.

While their deed may not seem extraordinary by Inuit standards, the Canadian military believe the Mamgarks' actions saved the life of one soldier and contributed to the safe return of an entire patrol.

Between Jan. 5 and 20, 1983, B Company of the First Battalion, the Royal Canadian Regiment, participated in Exercise Kovick. As part of the exercise, a seven-man patrol and two Ranger guides were dispatched on an overnight long range patrol to a lake about 20 km from the main company position here.

A blinding blizzard stranded the patrol on the fourth day out and by the sixth day, the men were running critically short of rations and fuel and were suffering from the cold. There was no communication with the main company. With winds gusting to 40 km and the temperature dipping below -30 Celsius, the decision was made to trek back to Eskimo Point. Led by the two rangers the patrol set out at 3 p.m. on two skidoos pulling komatiks.

During the trip back, a soldier was thrown from a sled and was struck by the trailing komatik. He was given emergency first aid by Warrant Officer D.P. Wood, placed in a sleeping bag and secured to Peter Mamgark's sled. After travelling 3 km. Johnny Mamgark's snowmobile broke down because of severe conditions.

Peter Mamgark, with two soldiers, continued at high speed to the Eskimo Point nursing station with the casualty, who was treated for a concussion, severe frostbite to his left hand, contusions to his ribcage, back and pelvis.

Johnny Mamgark had to guide the remainder of the patrol home on foot. They arrived without incident in total darkness with worsening weather conditions.

"Their courageous action undoubtedly saved the patrol from extreme hardship and possible serious injury as weather conditions worsened the next day," the commendation reads.

It was a proud moment for the community.

– 27 –

With a 1940s Rifle, He Stands on Guard for Thee

William Marsden

Montreal Gazette, 5 October 1985

Resolute Bay, High Arctic—On the shores of Resolute Bay, beyond a graveyard of rusting snowmobiles, Sgt. Walter Audla crouches low behind his Honda three-wheeler. His rifle rests against the seat, pointing down a shallow gravel gully toward the figure of a man outlined in black against a white target.

Behind Audla is Barrow Strait, where the U.S. icebreaker Polar Sea passed one month earlier without Canadian consent, sparking renewed political cries to protect Canada's sovereignty. Now, the red and white Canadian icebreaker Sir John A. Macdonald rests at anchor.

Audla, bracing against a sub-zero wind sweeping inland off the Northwest Passage, aims, fires and hits the target figure squarely in the chest. He's getting ready for the enemy.

In his uniform, a scarlet baseball cap and armband, Audla stands on guard for thee. Should the Red Army suddenly dash over the ice packs, he and his men – a military reserve unit of 15 Inuit called the Rangers – would probably fire the first shots in anger, if the world wasn't already nuked.

Outfitted with .303 bolt-action Lee-Enfield Second World War rifles and 200 rounds of ammunition a year, they are Canada's only permanent military presence in the High Arctic.

And that presence has recently taken on a new meaning here, where interest in the Rangers seems to rise and fall according to the latest U.S. ship going through the Northwest Passage.

When the oil tanker Manhattan went through the passage in 1969, Ranger patrols were reactivated by the department of national defence.

The Polar Sea has initiated a new flurry of recruiting and training sessions. At Resolute Bay, for instance, four new recruits were enrolled last month.

Casual Group

Audla, 47, is sergeant of the Resolute Bay outpost. It's a casual group of Inuit skilled in Arctic survival whose job is to report sightings of submarines and strange aircraft.

The Rangers would act as a modern version of the Indian scout if the regular army ever had to fight an arctic campaign.

Pockets of regular forces periodically conduct manoeuvers in the Arctic. But Canada's permanent armed force in the vast Arctic archipelago, with its million square kilometers of land and sea, is just two tiny Ranger units of about 15 men each, one in Resolute Bay and the other 383 kilometers northeast in Grise Fiord, the nation's two most northerly communities.

"The intent is to provide a military presence in the Arctic," Brig. Gen. Mark Dodd, commander of the Arctic Rangers, says from his base in Yellowknife.

"We have to do our part," Audla says, very seriously. "But it has its good part and bad part. Inuit don't like killing people."

The Rangers don't actually patrol as a group. They simply go about their normal lives of hunting and fishing and working around the settlement. If they see a submarine, for instance, they are expected to report the sighting to headquarters in Yellowknife, which would then dispatch a reconnaissance aircraft.

Reliable weapon

At a recent session with the Resolute Bay patrol, members were given photocopies of submarines on the surface – just so they would know what they looked like.

Three army regulars (one a British paratrooper) trained them to use radios and maps and how to fire the .303, which Dodd says is a "highly reliable weapon."

Throughout the Norwest Territories and Labrador, there are 36 Ranger patrols, with about 700 members. In the past few years, they have reported several aircraft sightings over Labrador. But the aircraft turned out to be West German planes flying out of Goose Bay.

There was one submarine sighting but it was never confirmed, Dodd says.

National Defence is preparing a white paper on defence policy. Dodd says he expects our military presence in the Arctic to expand.

"We will probably be giving (the Rangers) uniforms and build armories for a more permanent presence."

For the people who live in the North, the Polar Sea voyage was less a challenge to sovereignty – a war of words that was played out by the politicians – than another painful reminder of the South's ignorance and carelessness of the Arctic and its people.

"I get very annoyed when I think that Canadians think of this place as a no-man's-land," says Larry Audlaluk, deputy fire chief for Grise Fiord, a tiny settlement of 105 people on the southern end of Ellesmere Island – most of which essentially *is* a no-man's land.

"They took us up here 32 years ago and until this Polar Sea thing happens it's almost as if they forgot they put us up here."

Few Canadians know that, until 1953, Canadian Inuit did not live in the High Arctic. The only residents were those stationed at an air force base at

Resolute Bay (established in 1947 and since closed) and several RCMP "flag stations" (with special constables hired from Greenland) whose job was to counter the intrusion of Greenland Inuit crossing Smith Sound to hunt on Ellesmere Island.

In 1953, fearful of losing Arctic sovereignty, the federal government asked Inuit families from Port Harrison in northern Quebec to move 1,800 kilometers north to Resolute Bay and Grise Fiord. Six families volunteered to go.

A hospital ship left the families on gravel shores during the summer, with whatever possessions they could carry including caribou tents, some tea, flour and two boxes of ammunition.

More families moved north and today the 200 Inuit in Grise and Resolute are one of our most solid arguments for sovereignty over the High Arctic islands and surrounding waters, said Neil Faulkner, assistant deputy minister of northern affairs.

'Tremendous activity'

The Inuit have transformed the High Arctic from a "frontier into a homeland. There's tremendous activity up there."

– 28 –

Canadian Rangers: 'Eyes and Ears of the Forces...'

Dave Elms

***The Northern Pen*, 29 October 1985**

A group of men from varying walks of life from northern Newfoundland to Southern Labrador congregated in St. Anthony during the weekend. Together, they comprise a unique and important wing of the Canadian Forces, however, many people don't know they exist.

John Fennemore Sr. joined the Canadian Rangers seven years ago when his father, William, who had been a Ranger for many years, retired. The Rangers have always been low-key, but Fennemore hopes that will soon change.

"There was never much promotional work done, but that's beginning to change," he says. "We're now able to take part in training seminars; I've travelled to Corner Brook and Stephenville in the past couple of years. The Canadian Forces have provided a liaison officer who gives us a lot of help. And with the seminar here this weekend, it will certainly give the platoon a boost."

The St. Anthony man holds the rank of lieutenant and is the platoon leader for the St. Anthony area, which includes Cook's Harbour, Boat Harbour, and Wild Bight. Other platoon members present included Roddickton, Flowers Cove, Southern Labrador and the Quebec North Shore. Fennemore derives a great deal of pleasure from being a Canadian Ranger.

Eyes and Ears

"We are the eyes and ears of the (Canadian) forces," he emphasizes with pride. "I feel I'm a part of the forces, and it gives me a lift knowing that."

The role of the Rangers is a simple but integral one – maintain observations of the sea and air and report any suspicious or unusual activities. They must also carry out coast watches and report any air or sea distresses. The Rangers provide assistance to regular Canadian Forces troops who may be in their respective areas, and assist the RCMP on a local level. Any information that is collected regarding the discovery of enemy agents or saboteurs is immediately forwarded to the Canadian forces base in St. John's.

...

Expanded

... Fennemore notes that there are no special requirements needed should one become interested in joining the Rangers, however, there is a full contingent of members in St. Anthony at the present time.

"As long as you're 17 and up, in good health and don't hold any criminal records all you have to do is offer your name to one of us or get in contact with the CF base in St. John's," he says. "Our liaison officer is Capt. Jan Kwasniewski."

There are some 725 active Canadian Rangers in Canada today, and they range from fishermen to lawyers, Capt. Kwasniewski noted.

Fennemore adds that some 24 active Rangers exist. Four members in St. Anthony retired last year.

"It doesn't matter what type of person you are, but in this area fishermen seem to be popular because they make their living on the sea and are in close contact with any ships that may be on the go," he says. "Two fellows from Cook's Harbour operate the lighthouse on Belle Isle, so they're in a good spot for observing."

The act of observing comes natural, but weapons training is not something that happens every day. During the weekend, the 24 local Rangers headed for Cremaillere where they took advantage of testing their target shooting skills. As many discovered, practice will make perfect. Several of the Rangers did show signs of being skilled marksmen.

Lectures were carried out by Capt. Kwasniewski and a weapons display was demonstrated by three members of "A" Company Royal Newfoundland Regiment of Corner Brook at the fire hall.

"Hopefully, the exposure created by the seminar here this weekend will let people know that the Canadian Rangers do exist in their backyard and that we can play an important part in the safety of our land," Fennemore says.

– 29 –

Native Rangers Keep True North Free

Christopher Wren

Globe and Mail, 3 April 1986

Frobisher Bay, N.W.T. —If an enemy ever sweeps down to invade North America, it will have to contend with the Ikkidluak brothers and their bolt-action rifles.

Iola and Lucassie Ikkidluak live in Lake Harbour, an Inuit settlement on Baffin Island. They wear the red baseball caps and armbands of the Canadian Rangers, a force of mostly indigenous reservists who have been issued Lee-Enfield rifles dating back to the Second World War and told to keep their eyes open for anything suspicious.

Canada maintains no combat forces in the North, thought it has an electronic surveillance base on Ellesmere Island and a northern military headquarters at Yellowknife. The Special Service Force, a quick-reaction army brigade, has the mission of repelling a land invasion of Canada, but it is based in Ontario.

So the job of patrolling the thinly populated northern expanses and asserting Canada's sovereignty is left to the Rangers, who occasionally report sighting things like unidentified submarines as they roam around by snowmobile or boat.

One such submarine sighting took place six months ago in the Hudson Strait off Kangiqsujuaq in Northern Quebec. Sometimes a surveillance aircraft is sent up to check out the reports.

"Any guy knows the area generally within 300 miles of his community," said Sergeant Bruce Martin, a Canadian paratrooper who trains Ranger units across the North. "If a guy's out fishing and he sees something, he kind of reports it. It's not official, but they're out patrolling all the time."

The Canadian Rangers can be called up in emergencies but otherwise train in their spare time. They are not obliged to show up for duty if they don't want to, but as Joannepannee Kolola said in the local dialect, Inuktitut, "It's always fun going out with the guys, especially going out on the land."

Kolola and the Ikkidluak brothers belong to a Ranger patrol of seven Inuit from Lake Harbour who travelled 100 miles by snowmobile across the tundra of the Meta Incognita Peninsula to help a visiting detachment of French soldiers learn winter warfare in Baffin Island's severe Arctic climate.

Altogether, 640 Rangers have been enlisted in 37 communities across northern Canada, according to Captain John Todd, a regular soldier who commands the Rangers from Yellowknife. Another 600 or so Rangers serve in Newfoundland and Labrador.

The Rangers in the North are selected from among local hunters and trappers, and all but three are Inuit or Indian. Every Ranger, Todd said, "has to be familiar with native skills, know the land and be able to survive."

The concept started with the Pacific Coast Militia Rangers, a group formed in the Second World War to warn British Columbia in case of a Japanese invasion. The technology of warfare has since changed; Rangers haven't.

"A lot of people mock the fact that we issue Lee-Enfields to the Rangers," Capt. Todd said. "The fact is that it's a far more suitable weapon for where they are. It works very well in the North, with a minimum of moving parts."

These Rangers, who have hunted since adolescence, need no training in marksmanship. They are given three days of basic training in their home settlements, followed by four days out on the land.

They are taught to use a map, though they have little need for it. "Their navigational ability is absolutely uncanny," Sgt. Martin said.

They are also taught to operate a radio and the FN semi-automatic rifle. The Rangers from Lake Harbour agreed that the FN, Canada's standard-issue infantry weapon, was faster and more accurate than their old Lee-Enfields. But Lucassie Ikkidluak complained that after target practice with the FN, "my chin was swollen from shooting."

Sgt. Martin said it would be foolish for him to try to teach the Rangers how to survive outdoors. Besides, Iola and Lucassie Ikkidluak said, they find an igloo and caribou skins more comfortable than an army tent and sleeping bags.

But they have carried military rations on training exercises. "There used to be some chocolate bars in one of the rations," Lucassie Ikkidluak said. "That's the only thing I miss. But we prefer fresh meet."

Of all the training, the Ikkidluak brothers said, they liked the marching drills best.

– 30 –

The North's Own Canadian Rangers

Doug Holmes

Up Here, February/March 1987

The Canadian Army visited Coppermine in 1952 with a pile of rifles left over from the Second World War. They gave one to a young Jack Alonak and asked him to shoot at a couple of targets. His aim was not too bad so they asked him to serve as a Canadian Ranger.

Alonak is now 65 and still hunts and fishes and traps for a living. And he is still a Ranger, helping to assert Canadian sovereignty in the Arctic.

There are 360 Rangers in 37 northern communities and they are often the only Canadian military presence in the remote areas of the Arctic. Equipped with a 1940s Lee Enfield .303 calibre rifle, and issued 200 rounds of ammunition a year, a red baseball cap and a red armband, the Rangers are expected to keep their eyes open for "suspicious or unusual activities."

The federal government initiated the Ranger program on May 23, 1947. It was based on the World War II Pacific Coast Militia Rangers, the "coast watchers" who had the job of ensuring the enemy did not penetrate into British Columbia.

After the war, the Ranger concept expanded to include all the remote areas of Canada, including the Arctic. Yet with technological advances in communications, aircraft surveillance and radar stations, the Rangers' role became less crucial in the 1950s.

The Rangers are still criticized when pitted against modern technology. "For detection of cruise missiles, I don't think the Inuit Rangers would be that effective," a Yellowknife businessman once told a parliamentary committee on external affairs and national defence.

But the role of the Rangers has changed since 1970 when the defence department established its Northern Region Headquarters in Yellowknife. While the Rangers still watch out for foreign intruders, they are now also used as guides and Arctic survival instructors for regular army troops.

The Ranger program takes advantage of native northerners' traditional way of life—their ability to live and survive on the land, and their hunting instinct, says Captain Bob Gauthier, the officer in charge of the Rangers at the Yellowknife headquarters. Without the Rangers, regular troops would operate in the North "with great difficulty and a significant loss of effectiveness and perhaps even unnecessary great risk."

Tommy Ross is a long-time Ranger in the Mackenzie River Delta. He recalls a time when his patrol took several boat loads of troops from Inuvik, on the east side of the Delta, across to Aklavik, on the west side. One of the boats tipped over and the soldiers "got some good, practical experience on learning how to dry out in stormy weather."

"I like teaching people in the armed forces," says Ross. "They're young and energetic. They want to know how to make a fire with wet wood, how to build a snow house, how to set snares, how to survive in the bush."

The Rangers lend support to up to eight military manoeuvers a year, including adventure training on the Mackenzie River, Arctic survival exercises and winter warfare courses.

Alonak recalls a course hosted by the Coppermine Rangers one winter when 40 military personnel came north for a week of on-the-land survival training. The Inuit Rangers taught each soldier how to build an igloo and then made them sleep in it overnight "with no heat, no nothing," he says. Two soldiers at a time were left alone in the middle of the treeless tundra and in the morning they had to find, on foot, where the rest of the crew had set up camp. "We wanted to see if they could stand the cold," says Alonak. "They did all right."

Seeing the Inuit work while accepting the harsh Arctic climate as a part of their daily lives increases the confidence of soldiers from the south, says Major Steve Joudry, of Northern Region Headquarters. "Down south, if it hits -25 (degrees) you wonder if you should go out at all. Up here at -25, it's like a Chinook has come thorough."

With the figure increasing each year, about 1,000 Canadian troops on Arctic exercises will learn survival techniques from the Rangers during the winter of 1986-87. Also, every summer the Rangers of Pangnirtung, on Baffin Island, sponsor a cadet indoctrination course. The cadets receive instruction on seal hunting, whaling, fishing, skinning, traditional food preparation methods and soap stone carving.

The Rangers gained international attention when a British army patrol specializing in guerilla warfare tactics arrived in the Northwest Territories last winter for some Arctic training. In February 1987, Rangers in the Frobisher Bay area will support a survival exercise conducted jointly by a company from the French army and a Canadian regiment. Australia's aboriginal military reserves are also looking to the Canadian Rangers to exchange ideas and advice.

Ottawa is being urged by the Inuit Tapirisat of Canada, a political lobby group representing Canada's 25,000 Inuit, to enhance the role of the Rangers and give them greater recognition in the armed forces. The Ranger program is the one way Inuit can contribute to national defence, says Rhoda Innuksuk, the president of the organization. "The Inuit are the only ones with a real understanding of the Arctic environment," she says. "That knowledge should be used."

In a part of the country where the federal government spends $1.5 billion a year, the Ranger program has a $210,000 budget. It has been called the most cost-effective government program in the North.

As one way to increase recognition, the Department of National Defence is considering giving the Rangers a uniform of more substance than an armband and baseball cap. During two exercises this winter, one on Baffin Island and the other in the High Arctic, Rangers will be temporarily issued a parka, wind pants and white camouflage gear similar to the attire of the regular troops. This will allow the Rangers to integrate with the military personnel so they can fully participate in the exercises, says Major Joudry.

Further thought on issuing a regular uniform will depend on the Rangers' response to this winter's experiment. Many Rangers favor their own clothing, particularly some of the more traditional Inuit who still make and wear caribou skin parkas and pants.

Some say the clothing has an edge over some of the manufactured winter gear produced in the south. The caribou parka is worn with the fur on the inside. The sleeves of the parka are usually paired with mittens so the strands of guard fur on both articles of clothing match up to prevent the Arctic winds from blowing up one's sleeve.

Communication with some Rangers is a problem. Many of them do not speak English or French. There are seven aboriginal languages in the Northwest Territories, each with varying regional dialects. The language barrier is often a "formidable obstacle" in communicating, says Captain Gauthier.

A translation service for correspondence and instruction in Inuktitut (the language of most Inuit) is occasionally used, he says. "We have distributed newsletters in English and Inuktitut and we hope to eventually prepare basic training precis in the native languages."

Captain Gauthier says the military tries to respect the special cultural differences of the Rangers, 99 per cent of whom are natives, either Inuit or Dene. Because of this the military allows a certain amount of "flexibility" with orders and punctuality. The hours of the day are not always relevant in the land of the midnight sun. Children play in the streets at two o'clock in the morning, meetings rarely start on time and watches have no place on the traplines.

"Once I'm out on the land, I'm it," says Ranger Ross. "I don't care if they're a bunch of generals. If it's my job to keep 20 men alive, I'm it."

The six military personnel in Yellowknife who are responsible for the Rangers attend cultural awareness courses to help them understand the native way, but "that is not to say we accept their ways," says Captain Gauthier. "We certainly attempt to gradually acquaint them with basic military rules."

The initial training of Rangers is considerably more sophisticated than in the days when Alonak was handed a rifle and told to shoot. Now new Rangers undergo a 10-day training course where they are taught first aid, how to read maps and how to use rifles, pistols, radios and other military gear. They are also taught how to march and salute; and are given recognition training so they know what a foreign intruder looks like if they see one. More advanced training is offered to the Rangers every two or three years.

The military formalities are gradually being accepted by the usually causal Rangers, acknowledges Alonak. "They're doing pretty good, they're even getting used to saluting."

The Rangers have no obligation to train except in the event that they are placed on active service, by order of the prime minister and federal cabinet, or when called out in an emergency (war, invasion, riot or insurrection). Yet they never turn down an offer for further training, and some communities even ask for more.

The Rangers receive a daily wage while on training exercises, but for the most part it is strictly volunteer. Recruiting is the responsibility of the locally-appointed sergeant, the highest rank in each Ranger patrol. "Some of the young guys are kind of nervous about joining the army, but I tell them they won't be sending us anywhere, we're just protecting Coppermine," says Alonak, who has been sergeant of the Coppermine patrol since 1979.

In some communities there is a waiting list to join as the army prefers to keep the number of Rangers in each patrol to a manageable size of 20. Free bullets are a major incentive for hunters seeking to join. It is almost impossible to earn a living these days by hunting and trapping considering the costs of snowmobiles and sleds, gasoline, rifles and traps. Every little bit helps and, says Ross, "I'm not going to put my nose up at 200 rounds a year."

The defence department also has to deal with the occasional request for Rangers from northern communities currently without a patrol. Most recently a request from the Inuit community of Sanikiluaq, located on an island in Hudson Bay, was turned down because of the "low likelihood of military activity in the area." The military would rather devote its attention to more strategically important locations such as along the route of the Northwest Passage, says Major Joudry.

Most Rangers are proud to be part of the armed forces. "If I wasn't already working I would like to be in the army full-time," says Ross, who works Monday to Friday as a radio broadcaster at the CBC station in Inuvik. Other Rangers recently have taken on work ranging from deck hands on oil rigs to members of the NWT Legislative Assembly. They are now "weekend trappers." Therefore they are only part-time Rangers.

The Ranger program is threatened because an increasing number of natives are abandoning hunting and trapping to make an hourly wage. "Although it hasn't become an unmanageable problem, it does warrant serious concern," says Captain Gauthier. "As the traditional native lifestyle dies out, the usefulness of the Rangers diminishes."

The military does not enforce its retirement age of 60 for the Rangers. The oldest Ranger is 72. It is the older hunters who have the most experience on the land and the most skill in the traditional native way of life, and therefore make the best Rangers.

Tommy Ross, who doesn't know how old he is because he was born before records were kept in the NWT, reassures Canadians that the Rangers have them protected. "Say if a missile drops on our part of the country, whoever has the trapline it falls on, he'll know where to go find it."

– 31 –

Unique Sentries on Guard in Arctic

Dan Leger

Vancouver Sun, 19 October 1987

Iqaluit, N.W.T.—Far from the crowded military centres in the South, away from the bureaucracy and regimentation of the regular armed forces, a band of unique sentries keeps watch on Canada's North.

They are the Canadian Rangers, a mostly Inuit militia of 1,400 men. Formed into 37 patrols, they provide a first line of defence in the Arctic and along the coasts of Newfoundland and Labrador.

During a February military exercise on Baffin Island, the Rangers had a chance to operate with the regular forces and teach the southerners a few tricks about surviving in the Arctic winter.

The prime task of the Rangers is to be the eyes and ears of the military over the endless tracts of Canada's Arctic archipelago.

In times of crisis, the reasoning goes, submarines might surface in the North to launch missiles or to land infiltrators, or hostile powers might dispatch saboteurs to disrupt communications and early warning systems.

Hostile subs

Although the chances of actually spotting a hostile submarine on the surface in these frigid Arctic waters are slim, the Rangers remain vigilant.

One young Ranger, here recently for the joint army-air force exercise Lightning Strike 87, surprised reporters who asked whether he had ever sighted an enemy sub.

"Not around here," he replied casually.

At an outpost established for the manoeuvres on the outskirts of Iqaluit, formerly Frobisher Bay, Inuit Rangers built igloos for the troops.

The snow houses went up rapidly, with the two expert builders hacking chunks of snow into precisely fitting shapes while their force of assistants filled gaps.

One young Ranger lounged on musk-ox hides in a completed igloo, puffing a cigarette while a pair of camp stoves heated the structure to a temperature approaching comfort, despite the minus-40-degree chill outside.

He explained to a reporter how the ancient art of snow-house making was being lost by a younger generation more accustomed to southern-style accommodation.

Igloo-building

The hardest jobs in igloo-building are finding the right kind of snow, then getting the top layers to fit.

"Myself, I've never made one," he added.

His seniors, who were building the igloos, spoke only Inuktitut. Attempts to interview them through an interpreter brought only bemused smiles.

But the youthful soldiers from CFB Petawawa working with the Inuit, members of the army's elite Special Service Force, seemed to be communicating all right.

In the way of infantrymen everywhere, they had learned to curse in Inuktitut and were applying expletives to each other with gusto.

Army spokesmen said living with the Inuit hunters proved valuable.

Maj. Glenn MacDonald, of the 8th Canadian Hussars, commanded forward base elements of the Special Service Force taking part in Lightning Strike 87.

He said his men gained expertise in cold-weather survival, in tracking on the barren Baffin wastes and in such basic skills as finding food in the seemingly lifeless landscape.

Hunting caribou

One day, the Rangers took a patrol of soldiers on a caribou hunt. While the regular army is not authorized in peacetime to shoot caribou, the soldiers learned something of the skills Inuit hunters have used for millenia to capture their prey.

The Rangers also provided invaluable help during an air assault at Cape Dyer, on the east coast of Baffin Island just north of the Arctic Circle.

The Rangers, some of whom travelled over brutal terrain for three days to reach the exercise, teamed up with Pathfinders from the Special Service Force to prepare the area for the air drop.

The Rangers helped guide 300 paratroops into the drop zone and assemble patrols for their "assault" on a nearby DEW Line site that had been occupied by "Soviet infiltrators."

As one community spokesman said, there is more to being a Ranger than just a ballcap, an arm band and the 200 rounds of ammunition provided each year by the armed forces. They appreciate the contact with the regular forces.

"It shows that someone in the South still cares," he said.

– 32 –

1-800 Call First Line of Defence

Miro Cernetig

The Globe and Mail, 7 August 1991

If the conning tower of a foreign nuclear-powered submarine breaks through the ice covering of the Beaufort Sea, Earl Esau knows what to do to protect Canada's sovereignty over the Arctic Archipelago.

Without firing a shot, he will sling his vintage Lee Enfield rifle over a shoulder, head to the nearest telephone--not always an easy task on the tundra –and dial a 1-800 number to his military commanders to the south.

"They told us not to shoot," says Mr. Esau, a 31-year-old hunter from Sachs Harbour, the most northern hamlet in the Western Arctic.

"They've got bigger guns than we do, eh," Paul Reddi says, sipping a coffee in the local co-operative, the general store that feeds the approximately 200 people who eke out a living on this rocky shore of Banks Island, where musk ox and polar bear far outnumber human inhabitants.

Dialing a 1-800 number may not seem a sophisticated or overly reliable manner of defending the nation's ownership of the ice-filled channels separating the islands of the Arctic archipelago, waterways that the United States and Britain do not recognize as exclusively Canadian. It is certainly not a high-tech, sure fire method of monitoring the passage of nuclear-powered submarines that engage in a deadly game of hide-and-seek hundreds, and sometimes thousands, of metres under the ice.

But as members of the Canadian Rangers, the Inuit and Inuvialuit hunters who patrol the North on snowmobiles for the Canadian military for $55 a day and gasoline, Mr. Esau and Mr. Reddi are one of the few tangible signs that Canada is actively pursuing its claim at the top of the world.

Six years after the U.S. Coast Guard icebreaker Polar Sea sailed through the Northwest Passage without asking Canada's permission, sparking a rare diplomatic row between the two allies, the Canadian government has abandoned many of the expensive promises it made to enhance Canadian control of the ice-clogged waterways in the Arctic archipelago.

"We have not enforced the policy that was announced by this current government in 1985," says Harriet Critchley, one of the world's leading specialists on the Arctic. "Because of the attacks on these various programs we have undermined our sovereign claim to the Arctic."

The Polar 8 icebreaker and a fleet of nuclear-powered submarines are two of the well-known items scuttled by the Progressive Conservative government because of budget cuts. But also gone by the wayside are the northern-terrain vehicle designed for soldiers on Arctic missions and, according to one military source, the northern base that was promised for Nanisivik.

Canada has also to complete plans to put sonar devices on the ocean floor to monitor submarine traffic through the waters the country claims. It is also widely believed that the Soviet Union and United States, not Canada, have the region's best maps, another important way of showing ownership to an isolated region.

Although delay or cancellation of these and other high-priced items designed to enhance Canada's northern sovereignty has saved taxpayers billions of dollars, observers say the country has paid another price.

"The all terrain vehicle has been cancelled," Ms. Critchley says from her office at the University of Calgary. "So, that's it for mobilizing the army in the north.

"Our destroyers are aging and the new ships coming on stream have hulls that are too thin. They cannot ever go near ice—it could tear them to shreds."

And without the Polar 8, the Canadian Coast Guard has only Class 3 and Class 4 icebreakers that are inadequate to keep shipping lanes open year-round in the Arctic Ocean, something the Soviets routinely accomplish. "Our icebreakers can only go into parts of the Arctic during parts of the year," Ms. Critchley says.

On the diplomatic front, Canada's success has been somewhat more encouraging, though still limited.

Canadians are major players in scientific issues involving the Arctic, and federal agencies and departments are consulted or take part in scientific missions in the disputed region. Such consultation is seen as an admission of Canada's jurisdiction in the area and touted by government officials as a strong element in Canada's historical claim to the disputed region.

But other than the United States' desire not to aggravate Ottawa, there is no treaty or public agreement that would prevent a repeat of the incident in 1986 when three U.S. nuclear-powered submarines broke through the ice at the North Pole. The Pentagon's decision to release pictures of the event was widely interpreted as a move to reinforce the United States' claim that the waters stretching from the North Pole down into the archipelago are international and not exclusively Canadian.

Undaunted, Canada has attempted to strengthen its claim by drawing formal boundaries that encircle the archipelago on maps. However, a treaty signed two years ago that requires the United States to inform Canada when U.S. ships or icebreakers enter the archipelago does not apply to submarines or other military vessels, a spokesman for the Department of External Affairs says.

As well, the treaty appears to be interpreted differently by each side. Spokesmen for External Affairs say that Canada reserves the right to reject entry into the archipelago, while in the United States it is widely believed that Canada has agreed to routinely grant permission to enter the disputed waters.

Both countries maintain that the treaty does not enhance or detract from each other's legal case.

If the question of who owns the northern waters does reach an international court or tribunal, the Canadian Rangers are expected to be an important part of Canada's case. While other military efforts have been scaled back in recent years, the number of Canadian Ranger units above the Arctic Circle has increased to 52 from about 38 within the past five years.

Dressed in red sweat shirts and baseball caps, the Rangers will never be confused with full-time soldiers. Asked to drill with their rifles, Rangers in the community of Tuktoyaktuk giggle after giving a display that lacks the usual military precision. "I guess we should have looked at the instruction manual," Clarence Rufus confesses sheepishly.

But the Rangers, who military officials acknowledge know the land above the Arctic Circle better than anyone, can be a source of military intelligence. They are shown pictures of submarines and aircraft and asked to write down and draw diagrams of any they see. Although the Rangers' primary role is to be a kind of civil defence force that can carry out search and rescue as well as first aid in the remote communities, military officials acknowledge that the Rangers have spotted fallen satellites and submarines.

Mr. Rufus, who has taken to calling himself Sergeant of his Ranger squad, recounts an incident in which Rangers stumbled upon a submarine.

"One Ranger asked if it was a whale," Mr. Rufus says. "The other guy said, 'If that's a whale, its got a guy walking on its back.'

"He didn't report it right away," Mr. Rufus says, adding that there was no telephone nearby.

Despite the incongruous image of Inuit stomping the ice pack for submarines and armed with a rifle, 200 cartridges and a 1-800 number, the Canadian military makes no apology for the program, established in 1947 as a means of defending Canada's vast coastline.

"The program works like a bomb," says Brigadier-General Joseph Gollner, commander of Northern Region Headquarters. "Anywhere you want to go in the Arctic, pick up a phone, put in your quarter, dial the 1-800 number and you'll get your quarter back."

Still, there does seem to be a cultural gap that separates the almost exclusively native Rangers and their mostly white commanders.

Military news releases describe the Rangers as "visible proof of Canadian sovereignty" in the North. But many of the Rangers interviewed don't believe the Arctic region belongs to Canada and don't see themselves as tools in a turf war with the United States. The region should belong to everyone, they say.

"I never think of sovereignty," Mr. Rufus says. "It's none of my business."

Says Mr. Esau: "I think the land and the water should belong to the Inuvialuit. We were here first."

DND photo

Vaunted as masters of cover and concealment in their West Coast environments, the Pacific Coast Militia Rangers were a grassroots response to the Japanese threat to British Columbia and the Yukon during the Second World War.

DND photo

The wartime Rangers were masters of improvisation, using specialized skills from previous military experience and their civilian life in their preparations to defend their homeland.

The Ranger, January 1944

In the PCMR, veteran soldiers and sportsmen instructed younger members of their units. The most important subjects were musketry, map reading, fieldcraft and signalling, with most training left up to the local leaders.

DND photo

During early Cold War training exercises, Canadian Rangers taught army personnel from southern Canada how to survive and operate in the north. In this image, Inuk Ranger TooToo from Churchill, Manitoba relays information to army personnel in a Penguin during Exercise Bulldog II in 1954.

DND photo

A Canadian Ranger guard of honour inspected by Hon. Ralph Campney and Sergeant Sageakdok of the Frobisher Bay (Iqaluit) platoon.

DND photo

A Canadian Ranger teaches a soldier how to build an igloo at Resolute, March 1971.

DND photo

"Bannock Binds a Better Team." Canadian Rangers from Arctic Bay, Holman Island, and Lac La Martre, NWT, trained with soldiers of 3rd Battalion, Princess Patricia's Canadian Light Infantry, at Camp Antler near Yellowknife on exercise Nanook Ranger II in March 1974.

1 CRPG photo

Troops of 1RCR and Rangers from Rankin Inlet pose in front of their new house in 1985.

5 CRPG photo

Canadian Rangers on the Atlantic coast practicing their marksmanship skills with the Lee Enfield .303 rifle on an improvised range.

DND photo

Canadian Rangers on patrol in the NWT in the mid-1980s.

DND photo

Canadian Rangers from Grise Fiord in traditional dress looking at a map with Capt. G.R. Gauthier, a Ranger staff officer from Yellowknife, in 1988.

DND photo

Canadian Ranger Josephie Kiguktak fires his .303 Lee Enfield rifle as Ranger staff and Rangers look on, 1988.

1 CRPG photo

Passing the baton on Exercise Baton Ranger, 1992.

DND photo

Canadian Rangers on a boat patrol off the Atlantic Coast, 1992.

DND photo

Canadian Rangers John Akana and Andrew Anaktak put to good use their markmanship skills during training near Kugluktuk, Nunavut, in 2001.

1 CRPG photo

Canadian Rangers celebrate their trek to the magnetic north pole at the culmination of Operation Kigliqaqvik Ranger, April 2002.

– 33 –

Baton Ranger

Capt Michael Mietzner

Sentinel 1992/4

Somewhere between Tutoyaktuk and Paulatuk on the Northwest Territories coastline – about 2,000 kilometers north of Edmonton, Alta. – four Canadian Forces Rangers' snowmobiles towing fuel-laden *quamutiks*, the Inuit version of a sled, disappear in a howling early-March blizzard. With winds gusting to 100 kilometers an hour, the temperature plunges to a mind-numbing -95C. The brutal cold is too dangerous even for the Rangers, and they wisely decide to stop and wait out the storm.

The four Rangers are taking part in *Baton Ranger*, a unique exercise marking both the Canadian Rangers' 50th anniversary and Canada's 125th birthday. The challenge – to pass a baton from Ranger patrol to Ranger patrol from CFB Esquimalt B.C. all along the coast of Canada's three oceans to arrive in St. John's, Nfld., on July 1, the country's birthday.

The 12-sided baton, made of Canadian maple, symbolizes the 10 provinces and two territories. Red and green cord wound around each end reflect the Rangers' colours.

The idea for this out-of-the-ordinary relay was actually born in the North about one and a half years ago, with the hope of having as many as possible of the men and women who serve as Rangers touch the baton along the way. By the end of its journey, the baton will have been transported more than 15,000 kilometers by foot, aircraft, automobile, boat, canoe, snowmobile, dog sled, and even snowshoe.

The baton began its journey on Jan. 13, and has visited exotic sites like Zeballos, Kitkatla, Eagle Plains, Old Crow, Gjoa Haven, Kangiqsujuaq, Makkovik, Flower's Cove, Heart's Delight and Ferryland along the way.

But is was after crossing into the Northwest Territories that the most ambitious leg of the relay took place. When Sgt Mario Aubin, a Ranger instructor from Northern Region Headquarters in Yellowknife, and Ranger Sgt Simeoni Natsek, from Repulse Bay, were handed the baton just outside Aklavik, a six-week, 8,400 kilometre journey lay ahead of them.Up until then, the baton had been passed from patrol to patrol and, with the exception of some legs through the Yukon, no one person had accompanied it for any long period of time or distance.

The initial mission of Sgts Aubin and Natsek was two-fold – to hand the baton over to Sgt Aubin's boss by April 26 in Lake Harbour, a small community on Baffin Island, and to carry out a route reconnaissance along the North

Warning System's unmanned short range radar sties, situated roughly along the former Distant Early Warning line.

Equipped with Bombardier Canada donated Ski-Doos, as well as radios, fuel, food, maps and navigation equipment, including the Magellan Global Positioning Systems computer, the pair set off. As the two trekkers approached each community along the way, two Rangers from each guided them from one community to the next, following various NWS sites. The Ranger guides met the pair at predetermined sites with enough fuel and provisions to get to the next community. However, emergency fuel and rations were carried on each of the four sleds, should anyone from the group get separated.

Sgts Aubin and Natsek had been guided by Ranger MCpl Darrel Nasogaluak and Ranger Frank Rufus, both from Tuktoyaktuk, when they decided to wait out the storm. When they were forced to stop, however, they realized they had lost their rear guide. This is not difficult to do in whiteout conditions, when you can't even see the hand in front of your face. Fortunately, the individual showed up the next morning in time for breakfast, having spent the night in a snow cave. It was a real and timely reminder that the North can be inhospitable and unforgiving.

"You cannot fight with nature when you are out there," says Sgt Aubin. "You have to work with it. Otherwise you get hurt, or worse…die."

Working with nature means building igloos, snow caves or, when snow conditions are poor, setting up a tent. A skilled Inuk can build an igloo, capable of sleeping three or four people, in about an hour and a half. However, the art of igloo building, like many other traditional skills, is rapidly disappearing, says Sgt Natsek. What concerns him most is that the lives of his people may very well depend on the skills they are no longer learning.

When word is received in a northern community that a hunter is lost or caught in a storm, the community reacts in one of two ways. If the community is aware that the 'lost' individual knows how to build an igloo, there is usually no rush to begin the search: otherwise a search party is dispatched immediately, says Sgt Natsek.

As the Rangers carried the baton cross-country, fresh frozen arctic char, caribou, biscuits and bannock were their staple foods. The fish and caribou were eaten raw and frozen, allowing them to 'cook' in the stomach. According to the Inuit, this generates more body heat and keeps the travellers warmer than if they'd eaten cooked, warm food.

The constant vibration of the snowmobile, the need to replenish much-needed body fluids, and the opportunity to rest or wake up made stops for tea a frequent and most welcome break in the Rangers' routine.

Riders sometimes started to doze off after hours on their snowmobiles with only the hypnotic drone of the engine and the single track in the snow ahead to keep them company.

The varied terrain – flat, wind-swept snow surfaces, open water, rock and gravel, or two-storey-high broken-up sea ice – was a constant challenge. "With information from a northern readiness patrol carried out just days earlier by an *Aurora* aircraft, we were able to circumvent the worst of the bad areas of sea ice and stay on schedule," says Sgt Aubin.

"Thank God for modern technology," adds Sgt Natsek. "In my grandparents' days, they would have had to risk going through those areas or turn back."

The travellers were really quite a sight. Clothing varied from snowmobile suits and issue mukluks – worn when it "wasn't too cold or windy" – to pants made of raw scraped, but not yet tanned polar bear hide, a caribou *anorak* (parka) and caribou or sealskin *kamiks* (mukluks) lined with felt, worn when it "was really cold and windy."

During one of those nasty -95C days, Sgt Natsek spent a half-hour defrosting his helmet from his balaclava, which in turn was frozen to his beard. "I shaved it off after that," he says.

The eight-week journey of Sgts Aubin and Natsek came to a successful conclusion in the hamlet of Lake Harbour and a celebratory feast was held in Pangnirtung, a nearby community. The community hall was filled with people and freshly caught char, seal, caribou and polar bear. When one of the Rangers at the gathering was asked why he wasn't married, he answered that he was living common-law with a woman because she didn't know how to build an igloo. He added, "If you find a woman who knows how to build an igloo, marry her, quick!"

The logbook accompanying the baton on its historic journey across Canada is filled with comments from Rangers ecstatic about their role in the celebration trek. Whether their resting place for the night was a comfortable bed in someone's home or hotel, a hard cot in a trapper's cabin, the floor of a noisy unmanned radar site's maintenance building or an igloo, each and every one of them feel they have been part of a special event.

"I was born in an igloo on Lyon Inlet, 80 kilometres from Repulse Bay," says Sgt Natsek. "A generation ago, that would have been about as far as I would have travelled. I enjoy being able to meet fellow Rangers across the North as well as fellow Canadians. This is a great country and I am glad I have been able to see so much of it in this very special way."

– 34 –

Rangers East and West

2Lt James Simiana

Sentinel, 1992/4

"Their morale is something. I think the world of just about every one of them."

As Maritime Command's Ranger liaison officer in St. John's, Nfld., Captain Tony Lynch adds that enthusiasm among the Atlantic Rangers has "gone crazy" and that the platoons display "really good *esprit de corps*."

Listening to the Regular Force infantry officer talk about the Canadian Rangers on Canada's East Coast is akin to listening to a proud father.

Capt Lynch and his staff oversee the approximately 1,050 Rangers organized into 10 companies of 30 platoons throughout Newfoundland, Labrador and Quebec's lower north shore. These Rangers – in contrast to those in Northern Region – are approximately 90 per cent non-native and comprise teachers, pilots, fishermen, wildlife officers, even a justice of the peace.

And they're a capable lot, says Capt Lynch. Each platoon trains up to eight times per year in subjects that include general service knowledge, radio communications and weapons handling. Trained to function as separate units, each platoon also runs its own shooting competition. Naturally, how to conduct a recce patrol is a big part of their training and, since holding a recce competition last fall, Capt Lynch reports the Atlantic Rangers companies are "screaming for another."

The enthusiasm and relish the East Coast Rangers take to their task impresses Regular military units, whether it's the Royal Canadian Regiment training in Quebec or the U.S. Navy Seals in Labrador. The East Coast Rangers have proven themselves as capable as anyone on exercise up North and walked away with recent shooting competitions, as well.

"They can operate in any terrain and in any climate with their own equipment, and they do it with a big grin on their faces – and still ask to do more," says Capt Lynch.

Although the strength of the East Coast Rangers is slated to remain at its current level, their counterparts on Canada's West Coast are experiencing an ambitions renewal.

Indeed, it's out West that the Canadian Rangers originally began. That birth took place in 1942, with the creation of the Pacific Coast Militia Rangers as a paramilitary force established to help counter a possible Japanese military invasion. After the Second World War, the merit of having such forces observe

Canada's lengthy coastline was validated by the formal establishment – on May 23, 1947 – of the Canadian Rangers.

A quarter century later, formal command of the Canadian Rangers was taken up by Atlantic Region in Halifax and Northern Region Headquarters in Yellowknife, N.W.T.

The Rangers' recent revival on the Pacific Coast amounts to a re-birth. By 1978, they had been reduced to nil strength. Thanks to the recent increased emphasis on Canadian sovereignty, however, the Rangers on the West Coast – encompassing the province of British Columbia and under the control of Pacific Region – were re-established in October 1990.

The first patrol was started in the north of Vancouver Island in Port Hardy in February 1991, and subsequent patrols have been established in the coastal communities of Port Simpson, Kitkatla, Ucluelet, Tahsis, Port McNeill, Zeballos and Gold River. Others have recently been established in Terrace – inland from Prince Rupert – and in Kincolith. By the end of this year, 20 patrols with approximately 600 members will be in place.

Like their fellow Rangers on the Atlantic, the Pacific Rangers are enthusiastic –"just good folks," according to Major Ian Hay, the Pacific Region Ranger Liaison Officer in Esquimalt, B.C.

Bringing those people into the Ranger program – which is aiming for 1,000 members by 1995 –naturally takes some doing.

With many British Columbia communities tied into either the fishing or forestry industries, the Ranger training staff have to contact community mayor or band councils to determine when it's appropriate to visit, provide an information briefing and determine local interest. Notes Maj Hay, "I've yet to have a community come back and say, "We don't want this."

That degree of acceptance, he says, owes much to the fact that at one time, during the Second World War, the Pacific Coast Militia Rangers numbered 15,000 members, many of them still around today. That association remains and is particularly strong in some native communities, such as Port Simpson and Kincolith, where Ranger service has carried on down through families and where, says Maj Hay, "people take a lot of pride in being Rangers."

Because a number of these communities are isolated, the Ranger training staff often reach them by floatplane, following a Buffalo ride to Prince Rupert provided by 442 Transport and Rescue Squadron from CFB Comox.

Given that some of these communities lack banks, hotels or even a Royal Canadian Mounted Police presence, it's something of an adventure to establish a patrol. But once that's done, a lot of the training is done by the patrols themselves.

"They're outdoor people because Rangers have to be self-sufficient on the land, and if they're caught short on something, they're often taught by their peers," says Maj Hay. "Many of the Rangers are loggers, fishermen or ex-military, so each has something to offer. After they've been together for a while, everyone knows everyone else's job."

Still while patrols are largely self-taught, the training staff instructs patrol leaders in the control and operation of their units, and then takes the patrols into exercise training.

As with Rangers elsewhere, those on the West Coast are an integral part of their communities. Approximately one-quarter of the 34 members of the Port McNeill patrol have been training with 442 Squadron as search and rescue spotters. They also do volunteer work for their town's senior citizens' home and for local service clubs. "Port McNeill's mayor is a Ranger," says Maj Hay, "and he thinks it's the greatest thing going."

"These people take pride in being Rangers," he adds. "Everyone I've met here is a good Canadian just wanting to serve his country."

– 35 –

Out on the Land

2Lt James Simiana

Sentinel, 1992/4

Bruce Hikhaitok stoops to feel the snow inside the animal track. Finding it soft, he knows his quarry has just passed through this section of tundra near the Arctic's Coronation Gulf. Crouching down, stalking quietly, he motions to move along the hard-packed snow to avoid making any noise that might give our position away.

Standing beside this Ranger guide is Pte Chad McNamara, of B Company, 2nd Battalion, Princess Patricia's Canadian Light Infantry, based in Winnipeg. Pte McNamara has hunted deer before in his native Cape Breton, N.S. This, however, is the first time he's stalked caribou, so he pays close attention to his guide.

This learning experience is part of the Arctic survival training and support given by the Ranger patrol of Coppermine, N.W.T., to troops on Exercise *Reliant Nordic*. *Reliant Nordic* is primarily a sovereignty exercise, meant to develop a cadre of Canadian Forces members acclimatized and trained to operate in the Arctic. But with just five members with Arctic experience of the more than 100-strong company of Regular and Reserve Force personnel, being here above the Arctic Circle is also very much an adventure.

Out on the tundra or on the neighbouring sea ice in an arc radiating approximately 80 kilometers from Coppermine, B Company's three platoons are in the Rangers' backyard. Before the platoons return south, they will discover just what living in the Arctic involves and how to operate in this often inhospitable climate.

Such exercises occur about six times annually and involve Regular Force units, occasionally augmented by Militia members.

In the 10 days in mid-March that these 2PPCLI troops spent in the field, the Rangers, besides acting as guides, show the soldiers how to hunt, gut and skin caribou; where to channel through the ice to fish for arctic char, cod or trout; and how to carve out snow blocks for use as wind breaks or igloos. The troops come to appreciate the Rangers' ability to travel and navigate on the land in winter conditions, how to dress for the cold, and how to properly secure cargo to the Inuit version of a sled – the *qamutik*. They also, more than once, witness the Inuit mechanical aptitude for repairing a broken down snowmobile.

Up North, out on the land, these things are very necessary. And passing on this knowledge to the Canadian Forces is one of the reasons the Canadian Rangers exist.

A component of the Reserves, the Rangers are located from Newfoundland to British Columbia. In Canada's North, they are found in communities stretching from the Yukon Territory through to the peninsular areas of northern Quebec and into Labrador. They serve as a paramilitary presence during peacetime. Patrols report any unusual or suspicious activities and incidents within their areas and collect information that may assist the Canadian Forces.

Being a Ranger isn't entirely volunteer work. Rangers receive Reserve pay during local training exercises, ground reconnaissance and rescue missions, and when assisting as guides or survival instructors for CF exercises.

In the event of an emergency, their role expands to observing and/or containing any hostile forces until the arrival of Canadian police or military personnel.

By the 1950s, there were some 87 Ranger patrols in Northern Region. That was down to 37 patrols with 636 Rangers by 1986, but the northern Rangers have recently enjoyed a resurgence of interest, with the increased emphasis on the protection of Canadian sovereignty in the Arctic.

By establishing Ranger programs in new communities, the northern Ranger patrols have increased to 57, with 1,250 members. With continued expansion, 66 patrols 1,658 Rangers are expected by 1995.

Most of Northern Area Rangers are native Canadians — Inuit, Indian or Metis.

Some become Rangers to get the red Ranger armbands, ballcaps and sweaters, the obsolete but trademark *Lee-Enfield* .303 calibre rifle and the 200 rounds of ammunition they receive annually to practice marksmanship—an obvious attraction for an Inuk entitled to hunt caribou.

But most join for other reasons. Some want to find out what it's like to be part of a Ranger patrol. Others join because friends and family are or have been members of the local patrol. Many have earned their Canadian Forces Decoration. Some have been with the Rangers for 20 to 30 years.

Charlie Evaglok became a Ranger because he likes "travelling out on the land." His regular job is working at the Echo Bay gold mine 190 kilometers south-east of Coppermine. Now, for a few days, he's happy accompanying B Company's 5 Platoon to its camp by the Kugaryuak River east of Coppermine.

Bruce Hikhaitok, who enjoys playing practical jokes on the troops, says he's a Ranger because he's "crazy, I guess. No, it's fun and adventure travelling out on the land. I like it." He also likes his snowmobile and scouring the land for caribou. When he can't find caribou, he invariably circles around the troop-laden *qamutiks* or opens his *Polaris Indy* for some high-speed passes over the ocean ice, which look enviable from the back of a *qamutik* where we are freezing. Little wonder that he finds the troops "slow travellers."

Evaglok confirms that for some people in Coppermine, being a Ranger is considered an important job; he enjoys his Ranger status and considers it as something he's doing for the town of Coppermine. Hikhaitok says what he likes the most about being a Ranger is the opportunity "to meet new people every now and then."

George Hala and Charlie Hokanank, two Rangers who join 5 Platoon later in the exercise, consider their Ranger work as important as their regular job—driving Coppermine's water truck. As Rangers, they have just completed a 20-hour non-stop return mail delivery trip to the North Warning System's PIN 4 radar site—no mean feat on four wheels, let alone a snowmobile over the open ocean ice.

Their Ranger job is one they do because "it's different," says Hokanak, and they will continue to do it for as long as they enjoy it.

Now during *Reliant Nordic*, the Coppermine Rangers feel they are teaching B Company troops a few things. Equally important, the troops feel their contact with the Rangers is beneficial.

Cpl Kevin Kitzul, a reservist with The Royal Regina Rifles in Saskatchewan, calls the local Rangers intelligent. "A stupid person up here would freeze," he says. Impressed with the Inuits' ability to live off the land, Cpl Kitzul believes the Rangers are vital to the Canadian Forces' operational capability in the Arctic. "I doubt we could operate without the Rangers, because we're used to a warmer climate," he says. "They know where to operate, and what we can and can't do. I'm sure we could get the job done, but without them we are certainly not as efficient."

MCpl Darren Guitard, 2PPCLI agrees that the Rangers are worthwhile as a force that represents the Canadian Forces in the North. "They're useful up here because they've lived here," he says. Having lived in Inuvik, N.W.T. for six years, he feels living in the Arctic is mainly a matter of adjusting to different climatic conditions. "The Rangers are here to guide us. And, in return, we can teach them a little more discipline and how to work with Regular Force soldiers.

"For the Regular Force, Exercise *Reliant Nordic* is a good experience – one that shows people who've never lived up here what it's like, shows them that this is still Canada," adds MCpl Guitard. "And if by chance something happens up North, it's good that we've had this sort of experience."

The Rangers' laid-back approach to getting things done out on the tundra is generally appreciated. For Pte Heath Cook, a reservist with The North Saskatchewan Regiment in Saskatoon, that attitude seemed to help his Arctic indoctrination. "They always seem to keep our morale up a bit," he says, "especially when we have less to do."

Pte Roy Maceachern, 2 PPCLI, sees merit to more contact between the Canadian Forces and its northern Rangers. "If you keep that relationship going, it will become easier and easier to get things done up here. And subsequent trips by soldiers will make the Rangers more effective in terms of what we expect of them."

Increased contact and the opportunity to work together more often is an idea that meets favourably with the Coppermine Rangers. Ranger Sgt John Akana, who's headed up the local patrol for the past five years, feels *Reliant Nordic* is useful for both troops and Rangers.

"It's a good exercise. I think the Rangers learn something from the soldiers and the soldiers learn something from us. It's good being able to show them things like travelling out on the land, and how to hunt and skin caribou."

Sgt Akana, who regards the troops as "pretty good people," hopes the Canadian Forces will return next year to Coppermine.

"I'd like to see them up here every year because it's worth it for the Rangers," says Ranger MCpl Angus Havioyak, Sgt Akana's second-in-command. "And we try to do our best. I think we do a good job for the soldiers and for Canada."

– 36 –

Survival Skills Essential for Rangers

Canadian Press Newswire, 9 October 1994

Yellowknife—When the temperature drops below -50 C and an Arctic wind blows across the tundra, knowing how to pee without freezing sensitive parts is a necessary survival skill for a woman.

The trick is to move quickly - and grit your teeth, says Edith Giroux, a Dene from Fort Simpson, west of Yellowknife.

Her first winter exercise as a Canadian Ranger taught Giroux that bit of outdoor lore, along with some how-to's about erecting a lean-to and building a blood-warming fire.

"I was worried about that," Giroux says with a laugh, recalling her permafrost-sans-potty experience. "You have to take your snowsuit all off and everything."

Knowing not only how to survive but to actually work under daunting conditions is the bedrock of the Rangers, a reserve component of the Canadian Forces.

Many already know the land well and are used to joining search parties hastily called when a hunter disappears or travellers become lost.

As Rangers, they perform community work but also help to train regular Forces personnel in survival and search techniques, while improving their own skills in map-reading and first aid.

There's also a certain status reserved for Rangers in many northern settlements. Pio Kopak, an Inuk hunter and trapper from Repulse Bay in the eastern Arctic, took his father's place five years ago when the 66-year-old Ranger retired.

"This is very important, we don't do it for fun," says Kopak, 26. "We help with search and rescue. If anybody gets hurt, we have to help those people."

The pay isn't much - about $50 a day during infrequent formal exercises - but each Ranger is given a rifle and 200 rounds of ammunition for target practice and hunting.

They also get a functional, if not funky, uniform including a red sweater and baseball-style cap with a Rangers crest.

That kit impressed even the Queen - she praised a unit of Rangers on parade as looking "very smart" during her visit to the North in late August.

For a generation more familiar with the snowmobile than the dog sled, the Rangers also offer an important link to the knowledge of their elders.

"It's a chance for older Rangers to pass on some of their experience," says Capt. Paul Chura, a regular Forces officer in charge of the northern Rangers. And if there was ever a war, says the military, Rangers would assist authorities "in discovering, reporting and apprehending enemy agents or saboteurs."

But there's not much call for spy-catching in the North these days.

Rangers are mostly busy with their day-to-day lives, hunting for caribou to feed their families, saving for new snowmobiles and attending weekly patrol meetings that range in size from 10 to 20 men and women.

Even search and rescue operations are considered fairly routine for Rangers like Nauyaq Ugyuk, an Inuk from the eastern Arctic village of Spence Bay.

A five-year veteran, he co-ordinated his first rescue last spring when three adults and a child disappeared while travelling on foot between two settlements. Blizzard conditions complicated the search.

"We couldn't spot the people for two days, three days, and it went up to five days," says Ugyuk, 40, a compact man with a serene demeanor.

"There was lots of snow and for four days there was a white-out. You can't even see the snow when it's like that. But once it cleared up, they were found. They were OK.

"When it's like that, the best thing to do is wait till it stops snowing."

Giroux, a tall, thin woman in jeans and hiking boots, relishes the chance to leave her desk job at the Fort Simpson school and work outdoors with the Rangers.

"I like to spend a lot of free time out in the woods. I learned to hunt with my brothers as a kid and this gives me a chance to practise."

And, she adds, you can't beat the food, which comes fast and fresh. "We've been hunting caribou and we've even had beaver tail - it's fatty and kind of sticky.

"I guess it's finger-lickin' good."

– 37 –

In this Army, a Red Ball Cap is Regular Gear

Peter Moon

Globe and Mail, 14 November 1994

They are the only Canadian soldiers who do not have to pass a medical examination, are allowed to wear their hair as long as they like and have no compulsory retirement age.

The Canadian Rangers, part-time soldiers who perform both paid and unpaid duties in remote areas of the North and on the Pacific and Atlantic Coasts, have another distinction. They do not wear regular military uniforms.

Their uniform consists of an easily recognized red baseball cap, a red sweatshirt, an armband and a vintage .303-calibre Lee-Enfield rifle.

As Simon Hiqiniq, an Inuit hunter and leader of the 30-member Ranger patrol in the 1,000-resident community of Gjoa Haven, NWT, says: "Everybody here knows our cap. The rifle? It works perfect, in any kind of weather."

The Rangers' red baseball cap is known throughout the North, but in the South most Canadians have never heard of the force, even though they have been part of the Canadian military for 52 years. They began in 1942 as the Pacific Coast Militia Rangers, who watched the coast of British Columbia after the fall of Singapore and Hong Kong and the bombing of Pearl Harbor by the Japanese.

After the Second World War, they expanded across the North to help deal with the threat of the Cold War. They watched for strange aircraft and submarines (which, on occasion, they reported seeing), helped provide security at isolated radar warning sites and helped train regular soldiers in cold-weather warfare.

Apart from a small electronic listening post on Alert on Ellesmere Island and small staff groups in Whitehorse and Yellowknife, the Canadian Forces have no other troops on a permanent basis in the top third of Canada.

A recently completed parliamentary study of Canadian defence policy called the Rangers "an inexpensive and important element of the reserve forces" and suggested greater use be made of them.

Today there are 3,236 Rangers, the vast majority of them Indian, Métis or Inuit, in coastal and inland communities across the North. As members of the Canadian Forces, they establish Canadian sovereignty in the remote Arctic and are part of the military's northern early warning system. They help train regular soldiers by acting as guides and experts in living off the land.

"They are the only people who can start a snowmobile at 55 below zero using methods nobody down here has thought about," said Lieutenant-Colonel Jacques Bérubé at defence headquarters in Ottawa.

"They have an uncanny sense of direction. They can always take you from point A to point B. We don't really know how they do it… They teach us survival skills. They teach our soldiers tricks they would not think of."

In addition, the Rangers act as voluntary, organized groups who can play a vital role in remote civil emergencies, such as forest fires, plane crashes and searches for missing hunters and tourists.

The total budget for the Rangers is $3-million, most of which is military pay for seven to nine days of formal military training they receive each year. The training includes first aid, map reading, report writing, radio procedures and search-and-rescue exercises in the field. Their voluntary duties and other training are done without pay.

For example: the first Ranger patrol in the remote Cree settlement of Attawapiskat on the coast of James Bay in northern Ontario was organized by the Canadian Forces last March. Two months later the 30-member patrol helped remove the 1,300 residents of Attawapiskat by air to Moosonee after floods threatened their community. Half the patrol helped organize shelter for the dislocated in Moosonee and the other half guarded homes and belongings in Attawapiskat. The Rangers received no pay for their work.

An army study last year noted: "Since their inception, Rangers have consistently exceeded their recruiting targets. Most patrols have a waiting list of applicants. The patrols parade and train frequently and with considerable pride and enthusiasm. It is noteworthy that 30 per cent of all Rangers are female; [and are] predominantly from the First Nations. Rangers and Ranger patrols do real things: they improve their communities, they rescue lost people, they fight fires, they guide [military and civil authorities], they patrol remote areas and they contribute to [the training of] the Canadian Forces."

"We don't train these guys to take their .303s and charge across the ice floes of Hudson Bay to shoot at [Russian] Whiskey-class submarines," said Captain David Scandrett, an army officer who works with the Rangers. "That's not going to happen."

Instead, the patrol in Attawapiskat is holding an exercise in February to deal with the mock crash of a 44-passenger aircraft at the community's single gravel-runway airfield.

"We've never planned for a plane crash here before," said David Nakogee, the patrol leader and Nishnawbe-Aski Police officer. "What would we do if something like that happened? Now we are planning what to do. When something like missing hunters happened we just ran around and didn't know what to do. Right now all we have to do is say, 'The Rangers are here. We can call them out. They are organized and trained.'"

The normal age range for Rangers is 18 to 65 but "there are numerous stories of regular soldiers on sovereignty patrols in the Arctic who have been exhausted by 70-year-old Rangers, who have walked their butts off and outshot them and everything else," Capt. Scandrett said.

Besides performing useful military and civil emergency tasks, the Rangers are important role models for young people in aboriginal communities.

The Ranger patrol in Attawapiskat was organized this year after the band council sent a letter to the Canadian Forces. "The lack of activities in our community have resulted in many tragedies, including suicides, vandalism, drug and solvent abuse," the letter said. "Our people need hope; something to help them learn self-discipline and community involvement."

The Canadian Forces have responded to similar pleas from other native communities by starting a trial Junior Ranger program this year. Under the program, young natives receive instruction in such things as survival training and native dancing.

CITIZEN RANGERS

Ranger strengths in Canada

British Columbia	658
Alberta	22
Yukon	225
Northwest Territories	921
Manitoba	23
Ontario	45
Quebec	346
Nfld./Labrador	996
Total	3,236

Source: Canadian Forces

– 38 –

A Survival Camp in the Middle of a Snowy Desert: A First in the History of the Inuit Rangers

Monique Giguère

Le Soleil, Dimanche Magazine, 14 April 1996[1]

Lac Klotz, Nunavik—A herd of beige and white caribou run across the frozen tundra, as agilely and gracefully as if they had wings. Suddenly the Lac Klotz landing strip appears, marked by the Inuit Rangers. The pink cloud from a smoke grenade stands out against the snow, and a long line of snowmobiles stretches along the edge of the strip. From the window of the Twin Otter, they look like a column of ants, each with a mummy standing beside it, bundled from head to foot in a parka.

Lac Klotz, which lies at at 61°N latitude, is where 100 Inuit Rangers—who are somewhat like a Far North variation on army reservists—will meet. Four days ago, they left their homes in the 14 far-flung villages in the Hudson Bay and Ungava Bay areas to travel here. The event, a first in the 54-year history of the Rangers, was planned from afar by Land Force Quebec Area.

"The purpose of the exercise is to validate the training given to the Inuit in all of Nunavik, since responsibility for northern Quebec was transferred from the Northwest Territories to Quebec Area in 1993," explains Colonel Robert Chartrand, who is in charge of the Rangers in Quebec, speaking on board the Dash 8 flying us from Saint-Hubert to Kuujjuaq (Fort Chimo). He has seized the opportunity to inspect the troops.

"We need to make sure the patrols can function all over the territory and use maps and compasses accurately," he continues.

The Canadian Rangers are a paramilitary group whose activities are little known to most Quebecers. "We use them for many things," says liaison officer Marc Morin. "For search and rescue, to find people lost on the tundra. In the event of an ecological incident like a shipwreck, or a natural disaster like the Blanc-Sablon avalanche. For plane crashes—imagine a 747 going down in the Arctic with 300 people on board. You have to move fast before the survivors freeze to death. The Rangers are experts in how to survive in the Far North. They're the first ones we call on in an emergency."

The Canadian Army likes to say that the Rangers are the eyes and ears of the Canadian Forces in Canada's sparsely inhabited northern coastal regions. The military counts on them for many things, such as reporting any unusual or suspicious activity on the ground, passing on any information that could be useful to the Canadian Forces, and supporting Regular Force personnel on

1. Originally published in French as "Un Camp de Survie en Plein Désert Blanc."

their manœuvres in the North. During the Cold War, a Ranger detected a Soviet submarine off the coast of Labrador.

Lac Klotz

These 100 Rangers are at Lac Klotz to put their knowledge to the test. "The idea behind the Défi Nunavik [Nunavik challenge] is to bring together the patrols from all the villages in northern Quebec so that they can compare their techniques for surviving in a desert-like environment and learn from each other, the weakest from the strongest," says instructor Sergeant Mario Aubin. "It's the first big meeting of this kind in the half-century history of the Rangers. A major trip. It's an event I'd like to see repeated every year, but it will probably never happen again."

The project cost $100,000. And the Army is not compensating the Inuit for the use of their snowmobiles. It provides gas and pays them a modest $44 to $77 per day, depending on their rank: Ranger, Corporal or Sergeant.

When the "white men's giant bird" from Kuujjuaq finally landed at Lac Klotz on 3 April, after a 24 hour delay caused by a blizzard, the Inuit were already there.

From a distance, the grouping of igloos resembles a picture on a Christmas card. But seen close up, the reality is not so idyllic.

It's a real survival camp. The bloody hides of freshly killed caribou lie at the entrances to the snow houses. Antlers have been removed and lie everywhere. A wolf's head stares with its dead eyes. Ptarmigans' folded wings are frozen to their backs. Small but fleshy brook trout and Arctic char from the morning's fishing are glassy-eyed and stiff as boards.

Space has been set aside for a dining area. Caribou legs with the hoofs still attached provide raw meat to help the Rangers restore their energy. The men carve themselves strips of flesh directly from the dead, frozen limbs. The bones are thrown to the dogs.

"Raw meat keeps your strength up and helps your body fight the cold. That's why the Inuit eat it," explains Sergeant Aubin, who says he once devoured half a pound of raw bacon. Colonel Chartrand tells us that the word "Eskimo" is a somewhat pejorative name given to the Inuit by the Cree—it means "eater of raw meat."

In addition to hunting and fishing to ensure their survival, various activities were held during the four days at Lac Klotz: marksmanship exercises with .303 rifles, construction of the biggest igloo in the world—into which 107 people were able to fit easily—and a contest for the fastest igloo builder. Two of the snow houses collapsed, but finally Allupa Etok, from Kangirsuk, was declared the winner. He is a master at the art of cutting blocks of snow at the correct angles and, when he raised his arm in victory, his roof held.

But despite the energy and enthusiasm shown by the Inuit in carrying out all these chores, not all of them earned an A from the instructor. "It's been a success for the majority of the patrols, but a disaster for some villages," says Sergeant Aubin. "Two or three patrols were disorganized, their snowmobiles were not in good condition and they wound up getting lost on the way."

Even so, the Inuit Rangers are endurance personified in the barrens, where the cold can kill just as surely as their .303 Lee Enfields, the old World War II rifles issued to them by the Canadian Army.

A 740 km trek by snowmobile

At 5:30 in the morning on 4 April, the members of the Umiujaq patrol are waking up in their igloo.

Davidee Niviaxie lights the Coleman stove and puts on water to boil for making tea. He eats a piece of bannock. Condensation is forming inside the roof of the igloo and drops of water are falling on our sleeping bags. Daniel gets up and wipes away the drips. He and Jessie Aragutak smoke cigarettes and chat quietly.

The patrol from Umiujaq arrived at Lac Klotz two days ago. They made the longest trip to get here: 740 km, with a *komatik* (sled) tied to the back of the snowmobile to carry their equipment and provisions.

The team left their village on Hudson Bay on 28 March. The trip took five days. "The route from Umiujaq is really bad. Rough terrain, with lots of ridges and hollows. The trip home will be easier. It's downhill," says Daniel Niviaxie, taking another look at the roof. "It's not the right snow," he says. "Too soft. Sometimes the igloos collapse."

During its five-day trip, the patrol from Umiujaq built four igloos, half snow and half tent. The men hope to find them still standing on their way back. It would mean one less task for the patrol, which has to hunt in order to eat. They could bring only the bare minimum of provisions: biscuits, teabags and a few canned goods. They couldn't overload the *komatik*, which was already weighed down with gasoline, animal hides and cooking pots, as well as tools in case of a breakdown.

Daniel, 31, who attended community college in Cornwall, Ontario, says that during a trip to Kuujjuarapik (Poste-de-la-Baleine), he brought along his global positioning system (GPS) to help him find his way. "A GPS is a satellite navigation instrument that indicates our position, the distance we still have to go, and the time. It's really useful in the North. But on the way, the cold was so intense that the device froze. It wasn't much use to me."

Tamussie and his huskies

Tamussie Sivuarapik, 54, doesn't use a GPS. He came alone, travelling the 200 km from his home in Povungnituk to Lac Klotz on a dogsled. One night when he left his dog team to hunt caribou, he returned to find them gone. His 11 huskies had run off, taking everything with them, right down to Tamussie's mittens.

"No, I wasn't afraid to be alone in the midst of this vast space," he says. "I built an igloo with my bare hands. I prayed and I asked for my dogs to be returned to me. In the morning, they had come back."

Tamussie's story is all true, except for one detail. The "Great Spirit" that brought back his dog team was none other than Sergeant Mario Aubin. He had left from a neighbouring village and was worried because he had lost track of Tamussie, so Mario retraced his route, caught up with the runaway huskies and brought them back to their master.

Tamussie is a kind of shaman—the archetypal Inuk in constant communion with winter. And his winter lasts nine long months, from September to May.

God knows life is rough in this arid, icy land. Lying in the Umiujaq patrol's igloo with Jessie and the two Niviaxies, the entry tunnel blocked by a large chunk of hardened snow, I felt like an animal caught in a trap. But when I looked up, I saw the luminous night through the translucent roof of the igloo.

Kuujjuaq, a transportation hub

The Far North is blizzard county

Blizzards are a normal part of life in the Far North. The wind is glacial and unpredictable, driving the falling snow sideways, reducing visibility to zero and changing direction suddenly. When the weather is good in Kuujjuaq, on Ungava Bay, there's a snowstorm in Povungnituk, on Hudson Bay. On 2 and 3 April, fog and gusts of wind grounded the Canadian Forces Twin Otter, which finally landed 24 hours behind schedule at Lac Klotz, where Rangers from 14 villages in Nunavik had been gathering since 28 March.

"In the North, everything depends on the weather," Major Carlo De Ciccio said more than once to the team of journalists waiting at Kuujjuaq (Fort Chimo). We were powerless in the face of the elements. The flight commander, Steve Leeming, told us how, after taking off from Yellowknife in the Northwest Territories, he had to land at Rankin Inlet, on the western shore of Hudson Bay, to wait out the storm before continuing on to Povungnituk and Kuujjuaq.

Kuujjuaq, the gateway to Quebec's Arctic, has 1,300 inhabitants. It is the largest of the 14 Inuit villages in northern Quebec and also serves as the administrative centre, with its hospital, school, police station, hotel, general store and co-op. But it is a bleak place. There are only a few rows of modest V-roofed houses, sitting low to the ground in the snowy expanse.

Everything is prefabricated. The streets are not really roads, but wide snowmobile trails used by a handful of trucks and the occasional small group of children.

"Kuujjuaq is a supply centre, a transportation hub," say the people who live here. They don't mention the pervasive atmosphere of bleakness and the sense of isolation in this small community located not far from where the taiga gives way to the tundra.

"Here, we're at the treeline," Rangers liaison officer Captain Marc Morin had said while looking out the plane's window a few minutes before landing the day before. "Twenty miles north of here is the tundra." The spruce trees become sparser up here, and their growth is stunted. "Toothpicks … Quebec bonsai!" the crew members joked.

– 39 –

Corporate Canada Meets Canadian Rangers

Peter Moon

The Globe and Mail, 21 September 1996

It was definitely not the kind of board-room meeting that John Craig Eaton is used to.

About two dozen people milled around in the Coffee Shop, one of two small cafés in this isolated Cree community of 1,400 in James Bay, 1,560 kilometres north of Toronto. They were laughing and talking in English and Cree, smoking, drinking tea and soft drinks. Nobody was in a hurry.

The chairman of Eaton's of Canada Ltd. is noted for keeping a tight schedule in running the department store chain. But on a recent evening, wearing a Canadian Forces uniform, he was, as honorary colonel of the Canadian Rangers in Northern Ontario, in a different role from his customary one and in a very different environment and culture.

Things were not going according to schedule and he was casting frequent glances at his watch.

He was a guest at a feast that the local 26-member ranger patrol had organized to honour his first visit to Fort Albany First Nation.

But the geese were still cooking and the feast had to be served and consumed before he could go ahead with the business of awarding certificates to the patrol members for their help in moving the community out in May when the Albany River flooded its banks, and assisting with the election of a patrol leader and two master corporals. (The Rangers are the only element of the Canadian Forces that elects its leaders.)

Col. Eaton was learning that the rangers at Fort Albany, like many ranger patrols, tend to do things their own way.

The Canadian Rangers are part-time citizen soldiers who act as spotters, guides and trainers for the Canadian Forces in isolated northern areas and in remote communities on the Pacific and Atlantic coasts. There are about 3,200 of them and the vast majority are aboriginals.

As one ranger explained: "We're kind of like the militia down south. But different."

Their uniforms consist of a red baseball cap, a red sweatshirt, an armband and a vintage .303 calibre Lee-Enfield rifle.

Col. Eaton, 59, was in Fort Albany as part of a two-day trip to visit rangers in Northern Ontario. He holds a commercial pilot's licence and flew from Toronto at the controls of the Eaton's company plane, taking three members of the Canadian Forces permanent ranger staff at Canadian Forces Base Borden with him. He wore an air-force uniform because he is also honorary colonel of 400 Squadron of the air-force reserve.

The visit to Fort Albany got off to a rocky start because the community is in the middle of a power struggle. A dissident group challenged the election of a new chief and council and has been occupying part of the band offices. Extra police moved into the community after some property was damaged, people were threatened and a warning that visiting aircraft might be shot at. As a result, major portions of Col. Eaton's long-planned itinerary had to be cancelled at the last moment.

The feast eventually got under way and he was able to present the certificates for the patrol members' help in the spring. The elections were then held for a new patrol leader and two master corporals. Members submitted names and candidates were asked to step outside while the rangers voted, without discussion, with a show of hands. Nicole Reuben, a 25-year-old mother of three children and one of seven women in the unit, was elected the new leader.

"That was very interesting," Col. Eaton said in an interview afterward. "But in the aboriginal culture women have been chiefs and elders and leaders. So I guess it didn't surprise me that in this day and age they would elect a woman to lead them. What amazed me was how non-competitive the vote was. It was done very simply."

As he was leaving later in the evening to fly to Moosonee, another female ranger, Ruby Edwards, shook his hand and, to general laughter, said: "Glad to meet you. Make sure I get a discount when I shop at your store when I go outside now that you know how much everything costs up here."

Fort Albany has only one store in the community, where prices, because of freight charges, are considerably higher than in the south.

At Moose Factory, the Cree community in the Moose River opposite Moosonee, Col. Eaton met Lawrence Chum, leader of the 10-member patrol, which recently conducted a 31-day search to recover the body of a drowning victim. Mr. Chum took him on a tour of the reserve, introduced him to other rangers, and showed him how he smokes fish and game in the traditional teepee at the rear of his house.

At the Mushkegowuk Council, the regional government for seven Cree communities in the James Bay area, Col. Eaton and Major David Scandrett, commanding officer of the rangers in Ontario, discussed the need to repair a broken dike at Fort Albany to prevent flooding next spring that could threaten the community's hospital.

The federal Department of Indian Affairs and Northern Development has asked the Canadian Forces to provide emergency engineering assistance to repair the dike before the winter freeze-up. The only way to get heavy equipment into Fort Albany is by barge and air.

Ed Chilton, a Mushkegowuk Council official, said the recent establishment of ranger patrols has proved invaluable in local search-and-rescue operations. "In the past there was a delay in getting search and rescue organized for missing people," he said.

"There was a delay in getting the Canadian Forces organized, getting their planes in and that added to stress in the communities. With the rangers we have confidence that things are going to react much more quickly."

With ranger help, he said, three local communities were given specialized firefighting training this year by the Canadian Forces after three children died as a result of a house fire in Peawanuk in 1994.

Col. Eaton is also a member of the Canadian Forces Liaison council, which helps reservists get time off work to do military training without losing their annual vacation time.

– 40 –

Nunavik Rangers Honoured in Montreal

Jane George

Nunatsiaq News, 26 November 1999

The Canadian Rangers who rushed to Kangiqsualujjuaq's aid after the community's January 1 avalanche were honoured last weekend in Montreal.

Montreal—Last Saturday at the Farnham military base near Montreal, Major General Maurice Baril, chief of the defence staff and head of the Canadian Forces, honoured the second Canadian Rangers patrol group for their selfless dedication and professionalism in the aftermath of Kangiqsualujjuaq's January 1 avalanche.

Following the disaster, Nunavik Rangers rushed to provide assistance to the community.

Baril awarded the Nunavik Rangers a special unit commendation for their outstanding efforts.

"Immediately, and in many cases in spite of great personal tragedy, members of the second Rangers patrol group present in the village began rescue operations," Baril said in his presentation speech.

"Soon, they were joined by Rangers from other patrols across Nunavik. Together, they worked in support of local authorities to rescue survivors, secure the area, and assist with the funeral preparations for the victims of the avalanche."

Baril said the "rapid and efficient reaction" of the nearly 50 Rangers who responded to this disaster saved lives and alleviated the suffering of those affected by the avalanche.

"Without their dedication, the toll in human suffering would surely have been higher. The leadership and moral support the Rangers provided in the face of this crisis was invaluable," Baril said.

Sammy Unatweenuk, from Kangiqsualujjuaq's Ranger patrol, accepted the certificate on behalf of the 14 ranger patrols in Nunavik.

On the night of the avalanche, Unatweenuk, like others in Kangiqsualujjuaq, dug with his bare hands in the snow to uncover survivors and kept on digging right through the night until relief from Kuujjuaq's rangers and volunteer firefighters arrived.

"This tragedy which cost the lives of nine Inuit from Kangiqsualujjuaq opened people's eyes to the fact that the Rangers are valiant members of the Canadian Forces," read the commendation certificate.

In honor of the commendation, the Rangers' flag will be flown during next year at the Farnham base. The Nunavik rangers will be able to fly a special pennant in honour of the commendation. A medal also accompanied the award.

"We're very pleased," Mattiusi Tulugak from Puvirnituq told Nunatsiaq News. "We worked so hard — this is a special event."

The award ceremony was attended by dignitaries from Ottawa, Quebec City and Nunavik, including Makivik Corporation President Pita Aatami, Kativik Regional Government chairman Johnny Adams, Kangiqsualujjuaq Mayor Maggie Emudluk, Kuujjuaq Mayor Michael Gordon and Nunavik MP Guy St. Julien.

Long-time Kangiqsuaqjuaq ranger Charlie Arngak also received a special medal in honour of his 22 years of service.

As Canada's military reservists in Nunavik, Rangers are to establish a radio link with the Canadian Forces headquarters in Montreal in the event of any Y-2K breakdown in communications or during any other disaster.

"Their role is to be ready to respond, although we don't believe there will be an emergency," said Major Claude Archambault.

During the week in Farnham, Nunavik's Rangers met their counterparts from the Lower North Shore region of Quebec.

They attended a regimental dinner, visited a local branch of the Canadian Legion and also took time out to go shopping and attend a hockey game in Montreal.

– 41 –

Rangers Savoured Long-Overdue Recognition

Dan Davidson

Whitehorse Star, 15 March 2000

Dawson City—When Governor General Adrienne Clarkson toured some of the territory last week, part of her purpose was to present a special service award to long-serving members of the Canadian Rangers.

She was carrying out an initiative begun last October by her predecessor, Romeo LeBlanc, before he left office.

While the five branches of the Canadian Rangers are part of the Canadian military establishment, there has never, in the 50-year history of the Rangers, been a way for the Canadian Forces to honour these special members.

LeBlanc ended that by lobbying to establish the Ranger Bar addition to the Special Service Medal. The Ranger Bar-Special Service Medal was to be presented to those who had been members for four years and participated in a minimum of three Rangers' exercises.

Clarkson made the first presentations Feb. 14 to a select group of 17 Rangers from all across the nation at Rideau Hall in Ottawa.

For the rest, it was determined that she'd present them during her first national tour.

During her Yukon foray, she made presentations to Rangers in Old Crow, Dawson City, Mayo, Ross River and Whitehorse.

Yukon Rangers are part of the First Canadian Ranger Patrol Group, which takes in the Yukon, N.W.T., Nunavut, Alberta, Saskatchewan, northern British Columbia and northern Manitoba.

According to Lt. Mark Gough, public affairs officer with the Canadian Forces Northern Area Headquarters in Yellowknife, "The Rangers perform the duties of providing a military presence in support of Canadian sovereignty, including reporting unusual activities, collecting local data of significance to support military operations, providing local expertise, assistance and advice, as guides and advisors, in search and rescue activities."

Aside from their actual military duties, Gough was quick to note that the Rangers also perform a number of other ceremonial and community service functions.

He stressed the Dawson patrol's involvement with the Yukon Quest sled dog race over a number of years. As well, Rangers from all over the North had participated in the recent Trans-Canada Trail Relay, which began in Tuktoyaktuk Feb. 19.

Rangers escorted the water carriers from Tuk to Inuvik, Tsiigehtchic, Fort McPherson and Dawson, as well as south to Mayo, Pelly Crossing, Carmacks and Whitehorse.

In Dawson, the Junior Rangers were also involved, according to Sgt. John Mitchell. "Members of our own Junior Ranger Patrol escorted the dog teams out of Dawson on (their) way to Mayo and south," he said.

In their distinctive red-hooded sweaters and military paraphernalia, the Rangers are a highly-visible addition to most ceremonial events in Dawson.

What's not perhaps so readily understood is that the members are volunteers. While they're provided with the ball cap, sweatshirt, badges, and a No. 4 Lee Enfield rifle, most of their other equipment (cooking supplies, snow machines, shelter tools, etc.) is their own.

A Department of National Defence backgrounder document notes that "As a rule, Rangers are paid only when they undergo formal military training. They receive per diem compensation for the use of their personal equipment on exclusively military missions."

– 42 –

Exercise Roving Searcher – Tahsis & Zeballos Patrols

The Ranger (4 CRPG), Winter 2000

Exercise Roving Searcher took place October 27 - November 1, 1999, in the Zeballos area, located on the rugged west coast of northern Vancouver Island.

Zeballos comes from the name Cevalos, who was 2IC during Juan Perez's landings in the late 1700's. When Spain ceded her claim to the unoccupied territories on the west coast of North America in 1790, the British changed the name Cevalos to Zeballos. This area was important for its gold (which the Spanish mined extensively) and later for its rich iron deposits. Around the turn of the 19th century the area became a major fishing port with a large cannery.

Tahsis comes from the Mowachaht word "tashees" meaning passage. Hundreds of years before Perez and Captain Cook landed at Friendly Cove in Nootka Sound, the Mowachaht people used Tahsis as their winter quarters. The Nomash and Zeballos Rivers provide trade routes north and south-east, through which the early inhabitants conducted their commerce.

Tahsis and Zeballos are both located at the end of two large inlets that cut inland from the Kyoquot and Nootka Sounds. These waterways then wind their way through more than 25 kms of rugged and mountainous terrain that offers many small bays ideal for sheltering or hiding vessels.

Exercise Roving Searcher was to involve a channel search of a number of the bays and small inlets by boat, with each unit acting as a mobile observation post while keeping in radio contact with the others. The training was to familiarize Patrol members with the local area as well as practicing a search for another boat that was to hide in one of the many bays.

The exercise was conducted by WO [Pete] Malcolm and WO [Dan] Hryhoryshen, and was to prove to be anything but routine.

Prior to the exercise actually taking place, instructing staff will usually recce the operations area. However, no sooner had WO Malcolm and WO Hryhoryshen left the dock for their recce with Zeballos Patrol Leader Greg Brooks when they were called out on an actual rescue mission. An engineer for Rugged Mountain Contracting had been dropped off by helicopter on the beach of a bay located on Twin Island. Unexpected high tides later in the day made it impossible for the helicopter to land and the engineer was suddenly stranded without the necessary clothing and provisions to endure the night.

When asked how he felt about being rescued, engineer Bill Pederson said, "It was a God send. I was almost hypothermic." He had apparently told the helicopter pilot that the tide would be too high, but the pilot felt he would be able to land in the same area. After discovering that the engineer had been right, it was the helicopter pilot who sent the distress call that alerted the Rangers.

As always seems to happen in this area, the weather started out great but managed to turn ugly as the day of the exercise grew nearer. 36 hours before the exercise was to begin storm clouds moved in and it began to rain. Not just any rain, but rain in enormous buckets that lashed about in the wind with a fury. By morning it still had not let up. The Zeballos River had risen 3 feet over night and ocean waves the size of steam locomotives rolled in from the channel, rendering any water exercise out of the question.

Watching this rampage of nature from the comfort of our hotel dining room, one could only wonder what would have happened to our engineer friend had he been stranded on Twin Island on a day like this! Nothing would be able to fly in this weather.

It was beginning to look like the entire exercise was a washout, but WO Malcolm was not about to be put off by "a bit of rain." The channel might be too rough to venture out on but there were still plenty of woods to have an exercise in. Rangers might get wet, but at least on land they would not drown.

Formulating an alternate exercise in which a pilot of an F -18 is reported to have crashed in the vicinity, a second recce was required to find a suitable area for the search to take place. Using Patrol Leader Greg Brook's four-wheel drive truck, the better part of a day was spent bouncing along miles of logging roads - most of which went up mountain grades that at times looked almost impassable. At the lower elevations small lakes of run-off covered parts of the road - some of which were so deep that the truck almost looked like it was going to float.

After four hours of driving, a suitable area had yet to be found until Greg remembered an overgrown road that he used to have a trap-line along. Overgrown is something of an understatement and the bumper of the truck was soon pushing over some fairly big trees in order to make any headway along what little of the road remained.

This road eventually lead to the old Central Zeballos Mine, which had ceased operation in the late 1960's. Around the entrance of the mine lay numerous cases containing core samples, and off to one side was a large, discarded propane tank that looked like a fuel pod from an aircraft. The mine shaft was partially boarded over but there was enough of an opening through the planks to have a look inside with the aid of HQ's infra-red camcorder. The infrared's distance is about 50 feet and displays the image in shades of green. Light green showed the quartz-streaked walls of the cave and dark green revealed a large pipe along one side and what looked to be a lot of water on the ground. Beyond that it was pitch black.

This site proved to be an excellent alternate for the exercise, but there still existed the problem of how to bring the Tahsis Rangers here. The original plan was to pick them up by boat (a mere 45 minutes away), but with the storm still in effect the only way to get there now was by vehicle. This would require a lengthy road trip in excess of four hours duration - each way. There existed an old logging road that could cut this time in half, but judging by the condition of the logging roads already traveled, taking this route might prove be far more dangerous than going by boat.

With the exercise to begin in less than 12 hours it was decided to wait a few more hours to see if there would be any change in the weather.

Sure enough, in the early hours of the morning the storm broke and the wind-swept water of the bay was reduced to a mild chop.

This lucky break opened a window of opportunity and before daylight appeared two boats made a quick run to pick up the Tahsis Rangers. Although conditions looked favorable, the possibility still existed that the present calm could be nothing more than a prelude for another howler. It was therefore decided to stick to plan B and do the ground search instead of risking it out on the water for the whole day.

When the Tahsis Rangers arrived, a second set of orders were given, after which a convoy of trucks headed out to assemble near the road to the old Central Zeballos Mine. Following their maps and compasses, Rangers would use the road as an entry point, whereupon they would conduct roving area searches through the dense forest for any sign of a downed aircraft and its pilot. Directing staff had brought a parachute along to be used on the original exercise, and this would be placed in the bush to indicate the crash site.

Locating the parachute, however, was not at all easy. The forest is exceedingly dense and dark, not to mention the parachute was also olive drab in color. Only after several hours of intense searching did Rangers eventually come across it, hanging from the branch of a tree near a small creek. The pilot (a Ranger volunteer) was also recovered and evacuated back to the road where a vehicle was brought up to medivac him out of the area.

It was by this time mid afternoon and the Rangers took a well-deserved break. The rugged terrain had proven to be very difficult to hike through and lunch was well past due. Two cases of IMPs [rations] were soon devoured, after which Rangers relaxed before the long walk back to the starting position. Some of the Rangers took the time to explore the mine, but an old cave-in a few hundred feet inside prevented them from going any further. From this point the shaft apparently goes straight in for some distance where it is known to make a sharp right turn before continuing on into the bowels of the mountain.

At the debriefing conducted back at the starting position, WO Malcolm and WO Hryhoryshen gave the Rangers good marks for their accomplishments. A few pointers that WO Malcolm gave were to be very thorough on searches and not to overlook any minor detail. Task delegation was another reminder and to "not assume people are always going to do things." WO Hryhoryshen also made mention of task delegation and the need for more precise unit control. There was also the consideration that the exercise itself had been changed from what had been planned initially, although Rangers had been quick to adapt to their new mission and to accomplish their objective.

The weather continued to improve throughout the day, and by the time everyone returned to town the channel showed hardly a ripple. After hiking through miles of bush all day the Tahsis Rangers were glad to return home by boat and not have to endure a lengthy road trip.

The trip back to Tahsis offered an opportunity to see what the original exercise would have entailed. Many small islands dot the channel and the coast line has various nooks and crannies that offer numerous possible hideouts for renegade shipping. The area is also known for its marine hazards. Various reefs and rock formations lay just under the surface of the water, ready to break a propeller or rip a twenty-foot section out of a hull. It takes someone who has an intimate knowledge of these waters to navigate it safely and our Zeballos and Tahsis Rangers know this area better than most.

– 43 –

Canadian Forces Want Bigger Role for Nunavut's Rangers

Aaron Spitzer

Nunatsiaq News, 26 January 2001

A report done by the Canadian Forces' northern area headquarters says Nunavut's rangers should be used more to protect Canadian sovereignty in the Arctic.

Iqaluit—Nunavut's Rangers are ranging farther to protect Canadian sovereignty in the Arctic.

A report released last month by the Canadian Forces Northern Area headquarters in Yellowknife recommends that Rangers conduct regular exercises in the High Arctic to assert Canada's ownership of that area.

This year, squads of the volunteer soldiers will travel to the Arctic Archipelago to make at least two so-called "sovereignty patrols."

One, in April, will take a team of Grise Fiord Rangers to Alexandra Fiord on the east coast of Ellesmere Island. From there they will use snowmobiles to make forays out onto the land and sea ice.

Another patrol in June or July will be staged out of Mould Bay on Prince Patrick Island in the Northwest Territories. Participants in that exercise will be drawn either from Resolute Bay or from the community of Holman in the Northwest Territories.

"What they will do is cover some area, and they will show the flag and report any unusual activity," said Maj. Yves Laroche, commanding officer of the Rangers for the Canadian Forces Northern Area.

The patrols will likely involve between four and eight Rangers, and spend around a week out on the land.

This year's trips will follow up on a High Arctic sovereignty patrol conducted jointly by the Rangers and the RCMP on Ellesmere Island last March.

Though participants on that exercise were on the look-out for foreign hunters rumoured to be poaching polar bears in the Alexandra Fiord region, Laroche said they didn't encounter anything illegal.

According to the report, called the Arctic Capability Study 2000, the sovereignty patrols are part of a new push to insure Canada's ownership of the High Arctic remains unchallenged.

In sovereignty, as in anything, possession is nine-tenths of the law, said Maj. Bob Knight, the former head of Nunavut's Rangers.

"In purely legal terms, they're proving Canada's sovereignty over the territory they're travelling through simply through their presence," he said.

"If you claim that you have sovereignty over a certain area, yet you have never been there, then someone else could turn around and say, 'Is that really your sovereign territory?'"

It's not foreign military activity Canadian Forces officials are worried about.

What concerns them, according to the study, is "a scenario where Canada's claim to sovereignty is whittled away by foreign commercial activity."

Central to that concern is the Northwest Passage, where, the report suggests, the onset of global warming could bring a surge in shipping activity.

The passage is thousands of nautical miles shorter than the standard sea-route through the Panama Canal, and could be a profitable short-cut for transport companies if the polar sea ice thins out.

Canada considers the passage to be a domestic waterway, and demands that other nations secure permission before passing through it.

But other countries, including the United States, say the passage is an international channel. Canadian law, including Canadian environmental regulations, should not apply in the waterway, they say.

"To maintain sovereignty over Arctic waters, Canada must know at all times what vessels are in her waters," reads the report.

Several hundred Nunavut residents, mostly Inuit, serve as Canadian Rangers. They are organized into 25 community-based patrols, each headed by an elected sergeant.

Every year the Rangers undergo 11-day training sessions, where they practice first-aid, navigation and traditional survival skills.

– 44 –

Canadian Rangers Receive National Recognition

Bonnie McLean

The Labradorian, 19 March 2001

Ten members of the 5 Canadian Rangers Patrol Group in Happy Valley-Goose Bay have been presented with Special Service Medals. The Rangers were honoured at a gathering on March 13.

"The reason we are here this evening is to acknowledge the contribution of the Rangers nationally and present the Special Service Medals with Ranger bar," said Commanding Officer of the Rangers, Lt-Col [B.G.] Bailey of the Canadian Forces.

"This medal is to acknowledge those special contributions that weren't captured in any other specific way—to recognize and honour the contributions that the Rangers have made to Canadian Sovereignty and to Canada as a whole."

Those receiving the medals included Cpl. Richard Learning, Cpl. Jim Blake, Rgr. Sam Morris, Master Cpl. Wallace Lyall, Sgt. Wilf Stuckless, Cpl. Tony Holwell, Sgt. Wayne Massie, Rgr. Russel Finlay, Rgr. Clayton Lyall, Master Cpl. Ed Sampson, Rgr. Sterling Saunders and Cpl. Wendell Hamel.

The Canadian Rangers were formed to have a military presence where there normally wouldn't be one, such as the high Arctic. Although they occasionally help out the RCMP with rescue missions, the Rangers' main purpose is to help the military as a sub-component of the military reserve.

"We're trained in military protocol," said Lieut. Joe Anderson, who has been involved with the Rangers for 16 years. "Anytime the Allies come in and want some survival training, in the winter or summer, we supply it to them. We act a guides and scouts for the regular forces."

Members of the Rangers are trained by Ranger Instructors from 9 Wing Gander."

"We get six training weekends a year and they teach us everything—map and compass, winter and summer survival, patrolling and shooting," said Mr. Anderson. "We go to shooting competitions every year in Ottawa. I'm bragging a little bit, but I have some top shooters in my patrol. The year before last, Master Cpl. Tony Holwell claimed 'top shot' across Canada for the Rangers."

The Rangers get paid military pay, according to rank, and have to be ready to help out the military-when needed-at a moment's notice.

"If there's a plane crash, for instance, and the Allies needed some people to go and help surveillance the area, it would be the Rangers' task to do it," said Mr. Anderson. "They would give us a call and say, 'there's a plane down, would you be able to get your troops together within the hour?' And I'd say, 'yes sir' and we're gone—we'll pick up our rifles, ammunition and our kit bag and go."

Women Involved

Ranger Bernadette Anderson is Joe Anderson's daughter. She said she joined because her father wanted one of his children to carry on the Ranger tradition.

"Its enjoyable and you get to meet a lot of different people and do a lot of different things," said Ms. Anderson.

"We get together with other Ranger units in other communities in Labrador and do outings which consists of doing everything in the wilderness—more or less living off the land. You survey each one of the territories that you're on and do a lot of map, compass and [orienteering]."

Ms. Anderson is also one of the first women to join 5 Canadian Ranger Patrol Group.

"The Colonel came in here one day and said "how come there's no women in this patrol?" I said 'no one wants to join,' so I decided to put out some advertising for it and suddenly I had three women come up to me—my daughter, Bernadette, Judy Morns and Coreen Michelin," said Mr. Anderson.

"You can't turn them down," he said, adding that another woman also recently transferred from the Rangers group in Black Tickle.

Lt.-Col. Bailey said that the work the Rangers do with the Allied Forces is quite valued.

"It's something that the Forces recognize as being quite important," he said. "In fact, they've almost made it a policy for troops that are going out and living on the land to be escorted by some Rangers and we're working towards that now.

"It means for sure that you've done well and you've got an excellent reputation."

– 45 –

"Very Special Forces" Head to Fort Knox

Adrian Humphreys

National Post, 13 March 2002

More than 1,300 Canadian Forces reservists are engaged in a week-long intensive combat exercise at the U.S. Army's military training facility in Fort Knox, Ky., home to some of the most sophisticated battle simulators and some 5,500 U.S. troops.

The U.S. military base at Fort Knox can boast of its powerful M1 Abrams Main Battle Tank, the computer-controlled target ranges and the incredible artificial town - complete with underground sewer system - used for urban combat war games.

But only the Canadian Rangers can boast of regular polar bear hunts.

For the first time, eight members of the Canadian Rangers, a unique army reserve unit comprising aboriginals living in remote Arctic and sub-Arctic communities, are engaged in an international training operation.

The Rangers will also participate in a large-scale joint Canada-U.S. battle simulation on Friday, acting as partisans and guides, as part of Exercise Bold Venture.

The Rangers are the only military presence in much of Canada's remote northern territory and stand out dramatically from the other soldiers at Fort Knox.

The long braids on many of the Rangers contrasts sharply with the military-style haircuts typically found on the base. Then there is their more casual and practical uniform, a bright red, hooded pullover with the Rangers logo on the front - a rifle crossing an axe behind three red maple leaves - and a red baseball cap.

They shun the C-7, the modern automatic machine gun of the Canadian military, in favour of the bolt-action .303 Lee Enfield rifle, a weapon lacking the firepower of an assault gun but one that rarely jams at -50C and is ideal for hunting game to eat on patrol.

The Rangers arrived in Kentucky on the weekend accompanied by Stella Blackbird - a Cree elder and a traditional medicine woman instead of an army chaplain that more typically travels with troops - and carrying an enormous teepee.

"We have massively confused the Americans," says Major David Scandrett, commanding officer of the 3rd Canadian Ranger Patrol Group, a complement of 350 Rangers from Northern Ontario.

"We have a truck configured to carry teepee poles and I would argue that it is the only truck of its type in any military force in the world. It carries 35-foot teepee poles on a custom-built rack."

Yesterday, at the request of the U.S. Army, the Rangers set up one of their three teepees at the gates of the Fort Knox base, making their presence inescapable despite their small numbers.

Both the U.S. and Canadian troops are surprised when they encounter the unit's name. Their better-known namesake is the U.S. Army Rangers, an elite force among the world's best-trained fighting units.

"The American Rangers are Special Forces. Our Rangers are very special forces," quips a Canadian Forces officer.

For safety reasons during the intense training exercise, the Rangers' long hair must be tightly braided and tucked into jackets and under helmets and only the tops of their red pullovers poke out from standard-issue green fatigues.

They have also taken up the C-7 assault rifle for the operation.

Even with this transformation, however, Major Scandrett says questions constantly come from fellow army officers about his troops and he sees it as a chance to introduce the Rangers to both Canadian and U.S. forces.

"They look and see the Rangers insignia and 'Ranger' has a certain connotation in the United States Army and we have to explain who we are and what we are, and then they are even more intrigued by virtue of the fact that the Rangers here are all native people from the north and they have no comprehension of the environment our Rangers operate in," he says.

Master Corporal Matthew Gull, 26, from Peawanuck, Ont., a Cree community on a river mouth off the coast of Hudson Bay, is second-in-command of the Rangers in Fort Knox. His home community of 260 people is accessible only by air or boat.

Other Rangers at Fort Knox range in age from 23 to 40 and come from Sachigo Lake, Kitchenuhmaykoosib, Kashechewan, Constance Lake and Moose Factory, a community of 1,500 on the mouth of the Moose River. At 550 km due north of North Bay, Ont., Moose Factory is the southernmost community represented and the only one linked by rail or road.

For some of the Rangers, this is their first trip south of Timmins, Ont.

"Some of them are having a little culture shock," says Master Corp. Gull. "One just came out of his own reserve where it's 50 below and yesterday he was standing outside in his shorts. He said he's going to get home and freeze his butt off.

"I've been with the Rangers for about six years now, so I've trained with the C-7. But for some of the Rangers here, this is the first time they have handled the C-7, so I have to give them a brief how-to. The new weapon for me was the M1 Abrams tank. It was pretty exciting to see that fire."

Not many tanks in Peawanuck? "Not many roads," he says.

At Fort Knox, the Rangers tackled the overhead firing range on Monday. That is where soldiers crawl through an obstacle course at night, underneath barbed-wire fences, while live ammunition is fired over their heads.

They drove M1 tanks, have crewed a tank in the most sophisticated tank simulator in the world, and become the first Canadian Rangers to rappel from helicopters.

"It was a pretty new experience for me and it was pretty intense. At first we did a dry run and I thought, 'Ah, this is going to be nothing.' And then, when we did it for real - with explosives going off around you and bullets whizzing by your head - the adrenaline gets pumping," says Master Corp. Gull.

The intensity, he says, rivals that of facing down a polar bear, the largest land carnivore.

He has hunted and shot about eight polar bears. He drags the 1,500-pound carcasses onto his sled and heads off on the 150 km journey back to Peawanuck, where he presents it to his mother, who turns the fur into mittens.

While the Rangers may be at a distinct disadvantage in their advanced weapons training, when it comes to operating in the north, few do it better, says Major Scandrett.

"The Canadian Rangers are masters in their own homes, on their own lands, in terms of competence in the bush and waters of the north," he says. "It is different because of the context of what we are doing. If you brought any other soldier into the north, the Canadian Rangers would be the experts."

Would they kick the U.S. Army Rangers' butt, one-on-one in an Arctic face-off?

"I would think that is an apt description," Major Scandrett says with a laugh. "You can't touch these guys in the north. It is their land. They live it, they know it. They have that knowledge you only possess by doing it. That is what makes them unique."

Being unique has forced some adjustments to operating within the strict confines of a spit-and-polish U.S. Army base that houses both a U.S. Marine Corps and advanced armour training.

"We're spending a lot of time explaining what is going on and people are going out of their way to make that cultural accommodation. I'm very pleased with the response," he says.

Master Corp. Gull says the Rangers are having a marvellous time at Fort Knox, despite some yelling and the gruelling schedule.

If there is a complaint, he says, it is the food.

"It's not like what my mom cooks," he says.

"The food here is not that bad, it's just that some of (us) wouldn't mind some caribou meat right about now.

– 46 –

Ranger Pride: Nunavik patrols strut their stuff for visiting Brigadier-General

Nunatsiaq News, 15 March 2002

Akulivik—Canadian Rangers and Junior Rangers showed off their best spit-and-polish last weekend when Brigadier-General Marc Caron, the head of Quebec's armed forces, took a whirlwind trip through Nunavik, stopping in Inukjuak, Akulivik and Kuujjuaq.

More than 500 Nunavimmiut are either Canadian Rangers or participate in its more youthful version as Junior Rangers.

Last Friday evening, Inukjuak rangers of all ages jammed the airport terminal to greet the general and his fellow officers as they arrived.

Later, in the Innalik School gym, they put on a well-oiled parade and treated the military guests to demonstrations of throat singing and Inuit games. Nine new Canadian Ranger recruits organized a buffet of Qallunaat and country foods.

"You are a key component of the Canadian Forces in a remote part of the world," Gen. Caron told the gathering. "You also provide leadership for your community."

While in Inukjuak, Gen. Caron transferred the leadership of the patrol from Allie Ohaituk, who had been patrol sergeant for 10 years, to Simeonie Nalukturuk.

Nalukturuk took a moment to tell the general what the Canadian Ranger movement means to his patrol.

"We strive to be better than yesterday," Nalukturuk told the general.

Gen. Caron also handed out service medals and two special coins, one commemorating his command, which he gave to Inukjuak's mayor Shaomik Inukpuk, and another honouring the Canadian Rangers, which he presented to Ohaituk.

The municipality of Inukjuak presented him with a carving.

After the official ceremony, Gen. Caron met with the Rangers to talk about their concerns

The Junior Rangers, aged 12 to 18, were thrilled to meet with the general in the school's library.

"How old are you? Is it hard to be a brigadier-general? When did you become a general? Do you have children? Is it your first time in the North?" they asked.

Some had more practical questions — and wanted to know, for example, if they would receive better boots and pants.

The Canadian Rangers asked if the "wear-and-tear" payment they receive for using their snowmobiles and komatiks during Ranger exercises would be increased to $100 from $75.

The standard-issue 303 rifles are also aging. "Our rifles are older than most of us," said one man.

Gen. Caron explained the federal government is hesitant to arm Canadian Rangers with bigger and better rifles, since these weapons could potentially be involved in violent offenses.

The next morning, Charlie Nowkawalk took the general to visit the Canadian Rangers' qammaq and igloos on the outskirts of Inukjuak.

This winter camp, which is used by many members of the community, is close to the site where Inukjuak will host next summer's camp for Junior Rangers from July 9 to 16.

Four Junior Rangers from every community in Nunavik will attend this camp.

During his brief stay in Akulivik, Gen. Caron also reviewed the community's Ranger patrols and presented many service medals.

Akulivik's 30-member Canadian Ranger patrol involves a hefty percentage of the adult population, many of whom have been Rangers for years.

The thrown-back shoulders of young Junior Rangers standing at attention showed how seriously they took the general's inspection.

In Akulivik, Gen. Caron was also warmly welcomed, receiving a carving as well as a nassak with the community's name and emblem.

Akulivik resident Abraham Irqu, 72, was to have received a Canadian Decoration for 52 years of continual service as a Canadian Ranger — one of the very few times anyone in the Canadian Armed Forces has ever qualified for this award in Canada.

But Irqu had gone hunting. He was not in Akulivik to receive the CD-4 award, which must be personally handed over by a commanding officer.

– 47 –

True North Strong, Free Thanks to the Rangers: Arctic regiment celebrates its 60th anniversary

Canadian Press

***Toronto Star*, 11 April 2002**

Resolute, Nunavut—A thin red line will begin to stretch from this remote Arctic community toward the North Magnetic Pole tomorrow, celebrating a unique military institution and reinforcing Canadian sovereignty when it may be increasingly challenged.

The Canadian Rangers, in their traditional red pullovers and ball caps, will celebrate their 60th anniversary this week by travelling 850 kilometres past polar bears, through freezing cold and over shifting sea ice to the place on the Arctic Ocean where compasses are no longer reliable because of the strong magnetic forces.

"I'm looking forward to it," says Sgt. Paddy Aqiatusuk of Resolute, the guide for Operation Kigliqaqvik and a Ranger for 14 years.

Created in 1942 to keep watch on Canada's Pacific coast during World War II, the Rangers have evolved into the military's eyes and ears in the Arctic.

Now, nearly 4,000 Rangers from almost every northern community in seven provinces and three territories lend their local knowledge and on-the-land smarts to regular patrols throughout some of the most remote areas on Earth.

Armed with almost indestructible Lee-Enfield rifles and travelling on snow vehicles, they keep an eye out for foreign airplanes or ships and monitor unmanned radar stations. They're a big part of local search-and-rescue efforts.

And they also give Canada crucial ammunition in disputes about northern sovereignty.

Although no one disputes Canadian ownership over Arctic islands, the United States and some European states consider the Northwest Passage that runs between them to be international waters.

Canada argues that the sea ice is an extension of the land and Ranger patrols such as Operation Kigliqaqvik are intended to be demonstrations of that.

"It's all about sovereignty here," says Capt. Rick Regan, deputy commanding officer of 1st Canadian Ranger Patrol Group.

"The Rangers are the ones letting us know what's going on in our own backyard."

So far, sea ice has limited the usefulness of Arctic waterways.

But thinning ice cover may bring sovereignty disputes forward.

Tourists in icebreakers have been cruising the Northwest Passage since 1995 and many speculate commercial shipping will also increase.

"It does strengthen (our argument), as long as it's part of a cumulative picture," says Rob Huebert, a marine law specialist and a director of the University of Calgary's Centre of Military and Strategic Studies.

But more than that, he says the Rangers provide good role models and productive opportunities to communities that often can use both.

"The kids really look up to it," says Sgt. Darrel Klemmer of Tulita in the Northwest Territories, who has been a Canadian Ranger for five years.

"It gives them a positive role model."

Rangers are often leaders in their home communities and patrol leaders are elected by their comrades.

"You've got to be well respected to lead the patrol," Klemmer says.

Operation Kigliqaqvik, expected to take four or five days one way, brings together Rangers from 28 different northern communities from east to west, their largest operation ever. While there are huge differences between the forest and bush of the Mackenzie Valley and the treeless tundra of the high Arctic, Klemmer says the Rangers are finding lots in common.

"You get 30 different Rangers together and they'll have 60 different ways of doing the same thing.

"But we talk about our families and our communities and tell stories of the old ways. Everybody has hunting in common. Everybody likes to hunt."

Says Ranger John Mitchell of Dawson [City], Yukon: "The Rangers are one of the things that link the whole North."

They link northerners with the south, too.

"People don't realize how far we are from the nation's capital," he says.

"The Rangers make you feel more like you're a Canadian."

Joining the Rangers is an opportunity to serve the local community, says Aqiatusuk.

"I like to travel and I like to help people that need help," says Aqiatusuk, who has been involved with a number of search-and-rescue operations.

But Klemmer says it's also a chance to serve the nation.

"We don't want other people intruding on our land without us knowing about it.

"This is Canada. This belongs to us."

– 48 –

A Knights Tale: Smith's Harbour man praises Canadian Rangers

Coretta Stacey

The Nor'wester, 17 April 2002

Smith's Harbour—Ron Knight retired a decorated Canadian Ranger, and a proud member of one of Canada's best kept secrets.

The Rangers is a reserve unit of the Canadian Forces formed in 1942 to help patrol remote areas where the military's presence was low, in the event of a possible Japanese invasion.

At the time, they were called the Pacific Coast Militia Rangers. Men from all walks of life were dedicated to this effort and after the war was over, the military expanded this service to cover all areas of Canada.

Smith's Harbour resident Ronald Knight was a Ranger, dedicating 35 years to the Springdale section. He was commander of the section before his retirement in October 2000.

Mr. Knight thoroughly enjoyed his service, but grins when he said that devotion to the Rangers wasn't what enticed him to join initially.

"My cousin Cyril Pelley was First Captain. He was talking about the Rangers one day and told me I should join. He said they would give me a rifle and all the ammunition I wanted to fire. That is what made me think of joining."

After joining, and the troop began spending a lot of time in the woods on exercises, he began to enjoy it for what it was. Soon, the rifle and ammunition privileges were contained, but Mr. Knight enjoyed it to the fullest.

"This is the ideal thing for young people who enjoy the outdoors."

Mr. Knight said this is a true introduction to the military, and in the end you have a greater value for your country and the men who serve.

The Rangers have minimal duties in comparison to regular military personnel, but what they do is very important to the force. These men, and women, patrol the coastline collecting data which the military can use. They are the eyes on the shore watching out for suspicious activity, drug smuggling and unidentified aircraft.

They are called on patrol six-eight exercises a year, using weekend hours to do so. Participants are given basic training in map and compass, winter survival and firearm safety.

"You get a sense of appreciation being part of the Rangers, as the military use you and the things you do. You learn to appreciate the land and the environment. As much as I wanted to join this for the extended use of a .303 rifle, I don't even like to shoot a moose now," he stated.

Mr. Knight's greatest pride was when he was asked to participate in the 50 years celebration of the Rangers, and 125 years of being part of Canada. For this celebration, the Rangers had a 12-sided baton, representing the 10 provinces and two territories. Mr. Knight had the opportunity of being one of the Rangers who helped carry it across the country, doing his stretch in Newfoundland. He also had the opportunity to sign the log book accompanying it.

"I have so many memories. I really enjoyed my days as a Ranger. I was saddened to retire, but my time had come."

Mr. Knight was glad to see his son Paul take an interest in the Rangers. He had four sons, but Paul is the only one to walk in his father's shoes on patrol.

Mr. Knight would love to see his son go on to get all he can out of his service to the Canadian Forces.

"We received medals for our years of service, medals we are all proud of, yet there was no danger in the work we had to perform. When you think of all the men and women who had to die only to receive prestigious medals after their death. It's sad."

– 49 –

Polar Expedition Called a Boost to Sovereignty

Stephanie Waddell

Whitehorse Star, 26 April 2002

An expedition that took Canadian Rangers from across the North to the edge of the magnetic North Pole was about gaining a sense of sovereignty.

"It's always been a national thing, you know, to wave the Canadian flag here," Sgt. John Mitchell of Dawson City said in an interview Thursday from Resolute Bay, Iqaluit.

"And I think what everybody, that I'm getting a sense of, got was a sort of personal understanding of sovereignty as it pertains to them. You know, it became a personal thing."

Mitchell was one of the 30 Rangers on the snowmobiling trek to the magnetic North Pole, which left from Resolute Bay on April 9.

The trek, Kigliqaqvik Ranger (which means "place on the edge of known land") was to assert Canadian sovereignty over an area of the Arctic that is seldom visited.

"We got 40 per cent of Canada sitting up here north of 60; we've got more than our share of natural resources for the country here; we've got some of the unique flora and fauna that you can't find anywhere else in the world, and vast areas of some of the most beautiful country, and I think any country in the world would be proud to have any of the above and we got the whole works here," Mitchell said.

Beyond that, he continued, there are the people - some of the most unique, interesting and independent ones to be found in the world.

After departing Resolute on April 9, the group covered more than 800 km north in a week, ending on sea ice more than 10 km north of Ellef Ringnes Island and about two km south of an open stretch of Arctic sea.

The patrol members abandoned plans to camp on the ice for the night because they were unable to go around or over the water, and there was also concern that a gap might open behind the patrol. Instead, the expedition travelled south to Cape Isachsen and camped there.

When the group reached the most northern area on the trip, it was on the edge of the magnetic North Pole.

On April 18, Prime Minister Jean Chretien and National Defence Minister Art Eggleton called to congratulate the troops on their expedition.

The prime minister spoke with Mitchell as well as Warrant Officer Kevin Mulhern of Yellowknife, Sgt. Darrel Klemmer of Tulita, N.W.T., Sgt. Paddy Aqiatusuk of Resolute Bay and Ranger Deborah Iqaluk of Resolute Bay.

While Eggleton shared greetings with the Rangers, the defence minister also told the troop about the Canadian soldiers who died in Afghanistan in a "friendly fire" incident.

"We are happy for the Rangers and sad for our soldiers in Afghanistan," Eggleton said.

The Rangers' trek was the first time Rangers from across the North (1 Canadian Ranger Patrol Group) have come together in a combined operation.

There were four Yukoners, 10 Rangers from the Northwest Territories, 16 from Nunavut, one from northern Manitoba, one from northern B.C. and one from northern Alberta involved in the expedition. There were also five non-commissioned officers who worked out of the base camp.

On Sunday, a banquet will be hosted in the Rangers' honour.

– 50 –

Canadian Forces Putting "Eyes and Ears" in the North

Alisha Sims

***Nickel Belt News*, 13 May 2002**

Jim Clipping is living proof that one is never too old to learn new things.

The 58-year-old Tadoule Lake resident is Patrol Commander for the 29 member Canadian Rangers patrol set up in his community in March.

Canadian Rangers provide a military presence in remote and isolated areas of Canada that would not otherwise be covered by the regular elements of the Canadian Forces. Their role is to be the "eyes and ears of the north" for the Canadian Forces in support of Canadian sovereignty.

Formally established on the Pacific coast in 1947, the Canadian Rangers program was phased in and out a number of times before being revived again in the 1990's. There are now about 3,500 Canadian Rangers in 140 remote and isolated communities in every province and territory except Nova Scotia, Prince Edward Island and New Brunswick. In 2000, the program expanded into the prairie provinces with Canadian Ranger patrols being set up in Tadoule Lake in January, Lac Brochet in March and Gillam in April. A meeting to determine interest was held in Lynn Lake on Sunday.

Clipping, a former band chief and council member, current vice-president of the Hunters and Trappers Association of Tadoule Lake and a trapper for as long as he can remember, signed up to be a Canadian Ranger for the challenge and a chance to do something different.

"They say you never stop learning and I'm learning a lot. I like helping people and I see this as a way to keep active."

While the patrol is still new to Tadoule Lake, Clipping is already seeing the benefits, especially for the patrol members who average 30 years in age.

"The people enrolling like the idea. It offers a change for those living out in isolation. Also, being a part of the military creates a sense of responsibility, a sense of belonging."

Canadian Rangers are assigned tasks from reporting unusual activities to acting as guides and leaders for Canadian Forces members in the north, said Ranger instructor Wade Jones. He works out of the 4th Canadian Ranger Patrol Group, Manitoba Ranger detachment in Winnipeg.

"If a Ranger sees something in their community or the area they travel in that's unusual, or something they normally don't see, we ask them to contact us. It might be a strange plane that landed on a lake, a strange activity around a dam site or something like that."

When collecting data, Rangers record the amenities available in their community, such as local businesses and airport runway lengths. “If we decide to have a military exercise there, we’ll know what’s available,” Jones said.

Prior to establishing the three new patrols, letters were sent to each community describing the program. With the community’s approval, informational meetings were held and interested persons enrolled.

“This is a volunteer program,” said Jones. “A person can come and go as they wish. They have the opportunity to join at any time and can quit at any time.” Men and women over the age of 18 can become Canadian Rangers. A background check is done to ensure the candidate does not have any significant criminal problems. They must be in good enough health to carry out their duties. Any skill is considered useful, including experience as a bush pilot, guide or working with hydro electricity.

Each Canadian Ranger is issued a .303 Lee Enfield no. 4 rifle, 200 rounds of ammunition for marksmanship practice, training material and a vibrant red Ranger sweatshirt, ball cap and training vest.

Seven-day training sessions cover basic military skills, including mapping, survival and weapons handling.

Patrol Commanders head the patrol in each community. Patrol members vote to determine who their Patrol Commander will be. They are not appointed. Patrol Commander for the 22 Rangers in Gillam is John Watt. In Lac Brochet, the 29 Canadian Rangers chose Norman Denechezhe to be their Patrol Commander.

Denechezhe, 35, joined the Canadian Rangers to help not only his home community of Lac Brochet, but also others nearby.

The summer fishing guide has put in a lot of work as Canadian Ranger so far.

“I like doing the drills and am looking forward to getting the target range set up but I didn’t know we’d have to do public speaking,” he chuckled.

Denechezhe joined 12 other Canadian Rangers in Thompson last week for a leadership seminar. Public speaking was one of the lessons.

– 51 –

Women Rangers: A Perfect Fit

Najwa Asmar

The Maple Leaf, 2 July 2002

When she isn't dealing with administrative and educational matters as a high-school principal in her hometown of Inukjuak in Northern Quebec (1 200 people), Betsy Epoo seeks out adventure with the Rangers of the 2 Canadian Ranger Patrol Group (2 CRPG-Quebec), a commitment that allows her to contribute to her community's well-being in other ways and gives her an opportunity to serve her country.

"I like the challenge of the Rangers training, and I see no contradiction between my civilian life and my life as a Ranger, far from it."

Ms. Epoo is one of 505 women who serve with the Canadian Rangers, a figure which represents almost 13 percent of the total membership (3 850 members as of June 7, 2002) and slightly exceeds that found in today's CF, where women account for 11.9* percent of total strength.

Ms. Epoo is also in charge of organizing the training and activities of the local Junior Canadian Rangers, where she holds the rank of corporal. "I have already taken part in search and rescue operations with my patrol," she points out, hastening to add: "I can do whatever men do and I've already done it."

As Major Claudia Ferland, the national co-ordinator of the Canadian Rangers Program, makes clear, "The role of the women Rangers is a perfectly natural role, defined by the skills proper to it and the community's culture and values rather than by gender."

"Very little distinction is made between men and women during training. Later on, each patrol develops its own identity, and differences may emerge that reflect the traditions unique to each particular community.

Our organization respects this diversity," she notes.

To ensure such respect, Maj Ferland warns people against any categorization of female members of the Canadian Rangers. She advises instead that we stress the complementary nature of the two sexes' contributions to the organization. "I have known women Rangers who excelled in specific activities like hunting or shooting and who were highly prized for these skills, with no thought given to what sex they were."

In April of this year, Ranger Debbie Iqaluk (1 CRPG-Nunavut) was part of an expedition of the Canadian Rangers to the North Magnetic Pole. "It was

---

* According to figures provided by the Director of Military Gender Integration and Employment Equity Director and updated on June 1, 2002.

tough but it was great team work," she insists. "No difference between men and women, we did the same amount of work as they did."

In 1998, Sergeant Sally Bennett decided to enroll in the Powell River Patrol, following its creation in B.C. in late 1998 (4 CRPG). "I thought it was a good way to do things with my 18-year-old son, so both of us enrolled." Her son has since found other interests, but Sgt Bennett has remained a member, and her enthusiasm for the program continues to grow. "We are a kind of military entity within the community, a kind of link between these communities and the CF."

And in this area, she believes that women play a pivotal role. "Women help a lot with the community...It doesn't matter if there are some limits physically, which happens very rarely, they'll find something else for her...We have women Rangers in their 60's and they have their place in the community and in the Rangers."

– 52 –

Operation Polar Bear: Civilian and military personnel on sovereignty patrol encounter dense fog and majestic mammals on [Resolution] Island

Miriam Hill

***Nunatsiaq News*, 16 August 2002**

Resolution Island—Ranger Jetaloo Kakee motions a group up a rocky hill on Resolution Island, a small land mass off the southern tip of Baffin Island.

Half a dozen civilians taking part in exercise Narwhal Ranger, a sovereignty patrol helping to assert the Canadian military's control in Northern parts, have just been deposited on the island by the HMCS Goose Bay.

The Naval vessel leaves the bay broadcasting Newfoundland music from her speakers as she sails into the fog.

Six Canadian Rangers from Iqaluit and Kimmirut are on 24-hour polar bear patrol on an island frequented by the summering mammals. "Shhh... this way," Kakee says.

He leads up the side of an embankment overlooking the campsite, pausing and gesturing for the group to come closer.

Just below the ledge where Kakee stands is a sleeping polar bear. The animal isn't full-grown, probably only between 1,100 and 1,200 pounds, Ranger Matto Michael says.

Kakee makes a soft noise and the animal wakes up, turning its head toward the group above it. The bear rolls itself to its feet and trundles off down the far side of the hill toward small ice floes that have been pushed against the land.

The bear appears to take its time deciding which ice floe it will stand on. Kakee makes loud cooing noises and the bear looks up. Michael throws small stones toward it to try and scare it off.

The excitement over, the group makes its way back down to the campsite where Canadian Forces personnel are setting up tents. Large white canvas structures stand next to peaked green tents and a small bright yellow one. People mill about, waiting for food.

Boxes filled with army rations are opened, propane stoves are lit and within the hour the group's stomachs are full.

Part of the exercise is to test military communications capabilities on the island, and after breakfast about eight people brave the fog and light rain to begin the two-kilometre hike up the island's winding road to the airstrip, where a communications tent has been set up.

The weather is worse on higher ground and the rain pelts down as high winds blow the drops in all directions.

Two men unwrap a portable antenna and try to assemble it in the less-than-optimal conditions.

There aren't any telephone poles or high trees to attach wires to help transmit radio signals, Capt. Ryan Walker explains. The wires on the antenna attach to a radio sheltered in a tent where a man plays with the frequencies to try and reach someone.

"It's so we have a safety net if something happens here," Walker says. "Also, it's to exercise our interoperability." That means it shows that the Navy and the Canadian Forces can pass voice mail, e-mail or digital images over high-frequency radio. With a little patience and fiddling, the radio signal is heard and responded to by both the Iqaluit Coast Guard and one of the Navy ships that deposited troops on the island.

After the weather clears, the Rangers set up targets facing the bay for target practice. They are put through their paces, safety-checking their equipment before firing on the six wooden targets.

Ranger Dino Tikivik says the exercise has allowed a sharing of knowledge between arms of the Forces.

"I think they were surprised by our knowledge of animal behaviour and weather," he says. "I don't think we'll be going home today."

As the day progresses, it becomes evident that the planned airlift out will not occur due to the appalling visibility.

The light begins to dim and the group has eaten another meal of rations when the approximately 25 people tuck in for a night guarded by Rangers with .303-rifles.

The night is quiet and dawn breaks with more fog and mist.

Officers are soon in contact by satellite phone with the air force and Kenn Borek Air Ltd. trying to find a way off the island.

At mid-morning, people trundle toward another site near the bay where two bears are wandering the hillside across the water. Sharing binoculars and snapping photos as the weather starts to break, the mood of participants lifts with the cloud cover.

The airstrip has good visibility now and the surrounding landscape, which couldn't be seen the day before, presents itself.

Through the rolling mountains and blue sky, the charter plane lands to take the first shuttle of people back to Iqaluit.

– 53 –

Rangers Assist in Forest Fire Evacuation

Sgt Peter Moon

3 CRPG *Newsletter*, October 2002

Members of the Sandy Lake Canadian Ranger Patrol played a key role in helping to evacuate their community when it was threatened by a forest fire.

"They did an excellent job," said Warrant Officer Robert Smith, an instructor with 3rd Canadian Ranger Patrol Group at Canadian Forces Base Borden.

Several fires burning near the First Nation community in northwestern Ontario in July raised the risk of heavy smoke affecting residents, particularly the elderly, sick, children and those with breathing problems. On the advice of the head nurse, Chief Rusty Fiddler ordered an evacuation and asked for the assistance of the Canadian Ranger patrol.

The Rangers helped with the evacuation of more than 400 people on eight flights to the regional evacuation centre in Geraldton, where some remained for up to eight days before being allowed to return to Sandy Lake. The Rangers helped with the loading and unloading of the aircraft and with transporting the evacuees from the Geraldton airport to the evacuation centre.

Warrant Officer Smith flew to Geraldton with Warrant Officer Bruce Dunn, another instructor, to assist the Sandy Lake Rangers.

In Geraldton, the Rangers helped to provide security for the evacuees and worked with the local fire department on fire prevention.

"The eight members of the Sandy Lake patrol performed extremely well," Warrant Officer Smith wrote in his report. "They were 'invaluable' to the Geraldton Fire Chief, John J. Marino.

"The Rangers were the link from the community of Geraldton to the community of Sandy Lake... They reassured the elders, some of whom had never left Sandy Lake before.

"The Rangers also acted as interpreters between their people and the evacuation staff, ensuring information was passed back and forth, whether it was directions or status reports on what was happening back in Sandy Lake, or identifying the needs and concerns of their people."

Two of the evacuees gave birth to two babies while they were in Geraldton.

The eight Rangers in Geraldton were: Master Corporals Carson Fiddler and John Mawakeesik, and Rangers Glen Kakepetum, Danny Linklater, Elaine Linklater, Sandra Meekis, Judy Meekis and Linda Anishinabie.

– 54 –

The Canadian Ranger: A Staunch Citizen

Najwa Asmar

The Maple Leaf, 30 October 2002

Some are the mayor of their town. Others are entrepreneurs or association presidents. Others still are school principals, municipal officials or social workers. But they all wear with pride the cap and red sweater of the Canadian Rangers Corps, symbols of their double commitment. As citizens, they have a duty towards their country and as Rangers, members of one of the five groups of Canadian Rangers, they patrol the furthermost corners of this great land to protect Canada's sovereignty.

Sergeant Paulusi Novalinga (2 Canadian Rangers Patrol Group) has been a Canadian Ranger for 30 years. He was made Mayor of Puvirnituq, in northern Quebec, one year ago. He believes the two jobs fit perfectly with each other. "Being a Canadian Ranger makes my job as mayor a lot easier, and vice versa," he said.

In Mr. Novalinga's family, being a Ranger is an occupation that is passed on from father to son. His father was a Ranger and so is his oldest son. Family spirit and a sense of community are the qualities that best define the Canadian Rangers. "Our community is located 1 700 km from the closest town! Rangers help with search and rescue (at the request of the CF) and are there to respond to emergencies or crisis," he explained. "We also perform our primary task as Canadian Rangers, and that, in a nutshell, is to keep an eye out and to report any suspicious activity in our area to military authorities."

Corporal Millie Angnaluak, 1 CRPG, joined the Rangers in 1994, while she was attending courses at an adult education school. She was attracted first by the idea of patrolling such wide-open spaces and later, by the wealth of knowledge that Rangers possess. She has increased since then her involvement in and commitment to her community of Kugluktuk (Nunavut). While her job at the local airport keeps her very busy, she finds the time to look after the Canadian Junior Rangers and to represent her community at special events. "It is very clear to me that being here as Canadian Rangers has an extremely positive impact on our community," she explained. "Not only are we there when help is needed, but we provide a good example for young people to follow."

At age 58, Jim Clipping, vice-president of the local hunter and trapper association, became the commanding officer of the new Canadian Rangers Patrol established in his community of Tadoule Lake, Man., last March, as part of 4 CRPG. For this experienced trapper and former member of the band council, joining the Rangers was the natural thing to do. "I like helping people," he said. "It keeps me active. And being part of the military, gives me a sense of responsibility and of belonging."

Last August, Mr. Clipping and 13 other Rangers found safe and sound, a person who had been lost for two days in the woods near Lac Brochet, Man. "Our patrol is very new, but we've already seen how our being here is good for us and for the community," he said.

Such examples provided by all five CRPGs could fill volumes to illustrate the dedication and dual commitments of the Rangers to their community and the organization. Vern MacRoberts, a former carpenter and a member of 4 CRPG since the 1970s, has contributed much to social services in his community over the years. Corporal Betsy Epoo, 2 CRPG, is the principal of a secondary school in Inukjuak, who spends a lot of time with the Rangers. Whether mayor or simply a citizen, Canadian Rangers seem to have no difficulty at all discharging those dual responsibilities. Their enthusiasm for the job is unrivalled.

– 55 –

Port aux Basques Rangers Mark 60th Anniversary

John Zillman

The Western Star, 26 November 2002

A Ranger patrol here has passed a significant milestone in its history.

The patrol is the oldest in Eastern Canada and has just celebrated its 60th anniversary. Don Churchill, a sergeant with the Port aux Basques patrol, said he is often asked questions about the role and origin of the Canadian Rangers.

Churchill said the Rangers originated from the Pacific Coast Militia Rangers, which was a group created in 1942 to act as a coastal watch during the Second World War. The Rangers were introduced to Newfoundland after the war and were formally established as part of the reserve militia in 1947.

"The patrol has 28 members and covers the area from Cape Anguille to Couteau Bay, along the coast and inland," he said of the Port aux Baques patrol.

Churchill said the first person to be active and to introduce the Canadian Ranger program to the province was the late Fred Cox, who was the first and only person to hold the rank of major.

The Rangers currently have 3,600 members across the country.

"They perform their duties while working in their civilian occupations, operating alone or in patrols," said Churchill.

He said the Rangers operate in extreme climactic and geographic conditions.

"They patrol our mountains and coastline and are equally at home beside the ocean or in the tundra," he said.

The official motto of the Canadian Rangers is Vigilans which literally means the guardians and summarizes their mandate.

"Rangers provide a military presence in those sparsely settled coastal and isolated areas of Canada that could not be conveniently or economically covered by other elements of the military," Churchill said.

He said they also contribute to the activities of the Canadian Armed Forces by providing expertise about the local area and participating as volunteers in rescue missions in response to accidents or natural disasters.

"The Rangers also play an important role in their communities by serving as role models, especially for young people," he said.

In 1998, the Rangers were reorganized into five patrol groups to reflect the geographical mosaic of Canada.

"Patrol groups are organized in Northern Canada with 58 patrols. Quebec has 19 patrols and there are nine in northern Ontario. On the Pacific coast and the inland of British Columbia there are 27 patrols," he said.

Tasks assigned to Rangers include providing a military presence, reporting unusual activities, collecting local data in support of military operations and conducting surveillance.

Some significant examples of Ranger activities include reporting unidentified vessels within Canadian waters, participating as observers on Canada's west coast to counter illegal immigrants as well as responding to disaster situations such as aircraft crashes.

"The Canadian Ranger uniform is based on the concept that the Rangers, being outdoors people, are well equipped and able to survive in their environment," said Churchill.

All patrol groups conduct basic training, annual training and operational training. Rangers also receive patrol sustainment training which may include courses in first aid, flood and fire evacuation planning. They also receive training in search and rescue, major air disaster assistance, communications, firing exercises and setting up bivouac sites such as igloos or tents.

He said the Rangers have a tremendous impact on the lives of people in their local communities.

"There are a significant number of Rangers who hold leadership positions," he said.

Rangers are active community members who have a positive influence on their peers and are often held up as role models for youth in the community.

"Without the Canadian Rangers, it would be difficult for our military personnel to provide defence services such as humanitarian assistance to Canadians in many parts of this vast country," he said.

– 56 –

Quest Mushers Set to Follow the Usual Trail

Stephanie Waddell

Whitehorse Star, 5 February 2003

The Yukon Quest trail will likely take the same route it normally does on the Canadian side of the border.

"We had looked at another route out of Whitehorse, but fortunately, God must be a musher there," Sgt. John Mitchell with the Canadian Rangers in Dawson City said in an interview Monday.

The Rangers prepare the Canadian side of the Quest trail. Although this winter has seen unseasonably warm temperatures, Mitchell said, the Yukon River has frozen enough that the usual trail can be formed.

"Concurrent to that, the guys made it all the way through and into Braeburn, so that's fine into there," he said. "I mean, we do have low snow this year, but there's a two-inch base."

While overall it will be rough, it won't likely be too bad going into Braeburn because it's the Trans Canada Trail, Mitchell pointed out.

In Carmacks, the Rangers have cut an overland trail due to open water. Going into Carmacks, mushers travel on and off the Yukon River two or three times, and this will eliminate one of those sections, Mitchell said.

"The main change that I know about, at this time, will be that little section overland; and that's only two kilometres at the most," he said.

Mitchell doesn't think the trail's status will change much over the next week unless the temperature is drastically warm.

"If we get right up to 0 (C) for a few days, we may have a problem, but right now, it's not actually looking that bad on the route," he said.

The rough trail is broken from Pelly Crossing. The worst part of the trail appears to be the lower part of Scroggie Creek, he said.

"There's some rock faces there where some boulders came down and, of course, promptly froze themselves into the middle of the trail," Mitchell said. "It's kind of hard to do anything there."

Provided there's no major storm, the trail appears to be in good condition from there.

The Rangers are looking at the trail near the Alaska border. There's been traffic on the Yukon River from Forty Mile to Cassier Creek.

"We're only looking at a 10-mile (16.1-km) section of trail that's unknown, and I've got no doubt in my mind the guys will handle that," Mitchell said.

The race is scheduled to start in downtown Whitehorse at 1 p.m. Sunday.

– 57 –

Elders Essential to Rangers

Salim Karam

The Maple Leaf, 26 February 2003

For the 3 900 Canadian Rangers who provide a military presence in the sparsely populated regions of Canada's Far North, coastal areas and isolated settlements, the contribution of elders is essential. They serve as an example to younger generations and carry the memory of the often obscure culture we have only just begun to discover.

"If I have a cultural question, I can phone and she provides me with assistance and advice," says Major David Scandrett, commanding officer of 3 Canadian Rangers Patrol Group (3 CRPG), referring to the Group's elder, Stella Blackbird, of the Manitoba Cree community. Ms. Blackbird, who has six children and six grandchildren, was integrated into the Canadian Rangers as a special advisor.

The vast majority of Canadian Rangers are Aboriginals, 98 percent in the case of 3 CRPG. "It is important that we have cultural accommodation," Maj Scandrett explained, adding that elders are respected because of "their wisdom and insight." 3 CRPG is responsible for northern Ontario and each of its 13 patrols has its own elder.

"Elders are involved at the community level. They are important and respected members of their community. They are an example to others because of what they have demonstrated in the course of their lifetime or because of certain situations they have faced," said Maj Claude Archambault, second-in-command of 2 CRPG, Quebec.

"Elders are not part of the Rangers' structure and do not play a particular military role. They are merely an example, and are between 40 and 70 years old," said Maj Archambault. The elder attached to his group is from the Puvirnituq community, on Hudson Bay.

"Elders have a very important role to play, primarily within their own community. They can also be important to the Rangers because of their wisdom or the sound advice they can give on subjects that the Rangers teach to members and Junior Rangers," added Maj Archambault.

Rangers are active members of the community and are often cited as models for young people. There is no age limit for joining and their involvement with the Rangers can be a lifelong commitment. A Canadian Ranger in the Yukon and another in Nunavik (northern Quebec) have each served for more than 50 years and are currently CF members with the longest active service record.

"That is why we have no age limit. The Canadian Forces are looking for the cultural and spiritual knowledge, wisdom and influence that these elders contribute to the Canadian Rangers organization and to the Junior Rangers Program," said Maj Archambault, national co-ordinator of the Canadian Rangers and Canadian Junior Rangers.

– 58 –

Rangers Flying High with Canada Post

Jeff Bell

Victoria Times-Colonist, 15 March 2003

Canada's "invisible army" has been given the stamp of approval.

A newly minted postage stamp that celebrates the 60-year history of the Canadian Rangers was given its Victoria unveiling Friday in front of political and military officials, as well as representatives from the six volunteer Ranger patrols that serve Vancouver Island.

The Rangers date back to the Second World War, Victoria MP David Anderson told the gathering at the headquarters of the Four Canadian Ranger Patrol Group, the regular army unit that currently trains civilian volunteers.

"It was founded as a militia group to monitor the Pacific coast and the Interior of British Columbia for enemy action during the Second World War," Anderson said. "The Canadian Rangers have grown from that to be a national organization. They provide Canadian forces with local expertise and advice, they assist in search-and-rescue and they assist in humanitarian efforts."

At present, close to 4,000 men and women in 160 patrols across Canada volunteer their time to the Rangers, serving primarily in remote areas.

On Vancouver Island, six patrols are active in Port Hardy, Port McNeill, Zeballos, Gold River, Tahsis and Ucluelet. Stewart Shipley of the Port Hardy contingent, who was on hand for Friday's ceremony, said he spent his working career in the army before deciding to join the Rangers at age 60.

Twelve years later, he is still going strong and eager to share the benefits of his experience with other Rangers, who come from a wide variety of backgrounds.

"We have all sorts," he said. "Loggers, bakers, all types."

Shipley said Rangers fill an important role for regular military forces. "We are part of the military. We're the eyes and ears of the military."

Capt. Jim Miller, who leads the army's training program for Rangers in the west, said the decision by Canada Post to honour the group will provide a definite boost.

"For us, this is just great. I think in the long run this is going to be a wonderful opportunity to bring out to the people what the Rangers are."

Much of the training for Rangers from B.C. and other western provinces is done at Metchosin's Albert Head facility.

– 59 –

Canadian Rangers Save Missing Trapper's Life

Sgt Peter Moon

3 CRPG *Newsletter*, May 2003

An elderly trapper who was forced to spend two days in the bush without heat or food, in temperatures that dropped as low as -35ºC, has survived his ordeal after being rescued by a Canadian Ranger search team.

Paul Beardy, 72, left Bearskin Lake by snowmobile to go trapping and got into difficulties. He was forced to abandon his sled when he got bogged down in deep snow on an old, overgrown trail. He was reported missing when another trapper found his abandoned sled and two-day old tracks.

A quick air search failed to locate him and a three-member Canadian Ranger search team, aided by three community members, set out on snowmobiles at 10 p.m.

"We found him at 3 a.m. the next morning," said Ranger Sergeant Stuart Kaminawatamin. "He was suffering from severe frost bite of the hands and feet."

Mr. Beardy told his rescuers that he became disabled when he tried to turn his snowmobile around to turn back for Bearskin Lake. He had an ulcer condition and moving his snowmobile in a confined space in deep snow caused an internal injury that disabled him so severely that he was unable to move. He was able to light a fire the first night but spent the next two nights without any heat.

The rescue team wrapped him in blankets over a bed of spruce boughs to insulate him from the frozen ground, next to two quickly-lit fires, and gave him first aid and warm drinks.

The searchers used a satellite phone to tell the band office that they had found Mr. Beardy and that he would have to be airlifted out because of his physical condition. Ranger Master Corporal Charles Brown immediately began organizing a medivac airlift.

The searchers took Mr. Beardy to a nearby lake where a medivac plane picked him up four hours later and took him to Round Lake medical centre before he was airlifted to Thunder Bay. Mr. Beardy was operated on for his internal injury and both his legs were amputated below the knee.

"The professional conduct of the search party resulted in saving this man's life," Sgt. Kaminawatamin said. "I would like to commend and congratulate them on a job well done."

The rescue was the sixth successful search and rescue mission mounted by the Bearskin Lake Canadian Ranger patrol this winter. Earlier rescues, caused by slushy travelling conditions, which made snowmobile travel impossible, involved an overdue hunter and four overdue elderly trappers.

"I can say," said Captain Guy Ingram, operations officer for 3rd Canadian Ranger Patrol Group, which operates the Canadian Ranger programme in Northern Ontario, "that Bearskin's Canadian Rangers have put into practice what they have been taught. It's through their professionalism that once again another life has been saved by the Rangers of 3CRPG."

The search party consisted of Sgt. Kaminawatamin, Rangers Gordon Moskotaywenene and Levi Fox, and community members Lawson McKay, Jeffrey Fiddler and Spencer Morris.

– 60 –

Rangers Rescue Stranded Snowmobilers

Sarah Elizabeth Brown

Whitehorse Star, 12 January 2004

When Joe Girves' cousin perked his head up and said he heard a gunshot, Joe put it off to more hallucinations and lay back down in the snow.

Both men had been seeing and hearing things that weren't there for days.

The pair of Anchorage, Alaska, relatives were weak, severely dehydrated and frostbitten after seven days stranded in the bush about 60 kilometres north of Dawson City.

This morning, 19-year-old Brandon Girves was medivaced to a hospital in his hometown, where he's likely to have at least part of his feet amputated due to frostbite. Joe's been told to expect feeling back in his own numb feet in five to six weeks.

They'd set out on a snowmobile expedition from Tok, Alaska, to Dawson on Jan. 2, only to have one of their machines break down just hours into the trip.

The Rangers, out on their annual winter training exercise, stumbled upon the stranded Alaskans on Saturday and promptly set up camp and administered first aid.

Because of the younger Girves' condition, the two men couldn't be transported out of the Clinton Creek area they'd been in for the past seven days. Instead, the Rangers took care of the pair for a day in the bush before a helicopter could transport them to the Dawson nursing station.

Ironically, just two hours after the Rangers inadvertently rescued the pair, Joe's girlfriend flew overhead in a search aircraft. When he'd missed his scheduled Jan. 7 phone call home, she'd notified the RCMP and Alaska State Troopers, who were grounded due to poor weather.

His girlfriend Francine had been flying in the chartered airplane for two days when she spotted the abandoned snowmachine, said Joe, 35, but it would have been another day until a ground crew could reach them.

"By then I don't know if Brandon would be alive or not," Joe said this morning as he polished off a hamburger and fries at the Walmart McDonald's.

"Those guys are real heroes, real professional gentlemen," said Joe about the couple of dozen Rangers who rescued them. "Babied us along, wouldn't let us do anything for ourselves."

Once Joe was flown to Whitehorse from Dawson, a local ambulance attendant dropped him off at the department store because he only had the clothes he'd been wearing for the last week.

He's currently headed to Vancouver to buy a ticket "to the nearest warm beach," Joe said this morning. He's going to take it easy for a while before going home to Alaska.

...

"It's all kind of a blur now," he said. Throughout the entire ordeal, the men were able to catch only one or two hours of sleep at a time, and it wasn't restful sleep, he said. He slept for five hours once the Rangers found them, but was up all last night organizing his cousin's medivac.

At their campsite, the Polish sausage and box of snack bars they'd brought with them on the second machine was gone in 24 hours. Joe had expected to be in Dawson buying snowmobile parts, not stranded in the bush.

The first several days were frigid, with temperatures in the -45 to -48 range, not including the windchill. They both sported several layers of long underwear, with fleece and snowsuits on top of that. Both men wore heavy winter boots and had several hats and layers of gloves.

They weren't able to get their winter survival tent close enough to the fire, so Joe cut it up to use as a blanket. One man would huddle on one side of the fire, holding a wool blanket around his shoulder to trap in the flames' warmth, while the other did the same with the tent on the other side. They both now have some level of smoke inhalation, said Joe, but they kept somewhat warm.

For the first three days the cousins had lots of energy to collect firewood and melt snow in their pan for drinking water.

But by the fourth day, the energy and their hunger pangs vanished.

...

By Jan. 8, they knew someone would have realized they were in trouble, but became more despondent as help didn't arrive that morning or the next.

At one point they moved their camp 100 metres to be closer to firewood, but it was "pathetic" watching themselves try to move. Despite endlessly melting water, both were seriously dehydrated.

For the last two days alone, the cousins alternated curling up together to keep warm between making the effort to build a fire.

"The last day was real bleak," said Joe.

Both men had been hallucinating for a while, and that's all Joe thought was happening when Brandon heard a gunshot the morning of Jan. 10. But a steady stream of 24 to 25 snowmachines pulled by, ridden by Canadian Rangers who'd seen their helmets the men had propped up in hopes of being spotted.

"I thought it was a hallucination," said Joe.

The Rangers immediately swung into action, setting up a tent for the exhausted pair and calling the Dawson nursing station via satellite phone for instructions about thawing out Brandon's feet.

It was then the cousins learned that poor weather was grounding rescue aircraft, said Joe. It wasn't until the next day that a helicopter could be flown in for a pickup.

They threw up the food the Rangers first gave them, said Joe, and from there on, the pair were given a regimen of fluids, then soup, then solids.

Brandon, who works in a coffee shop back home, went from 113 kilograms (250 pounds) to 86 kg (190 pounds), while Joe lost about 14 kg (30 pounds).

The Rangers told them they'd not met many people who survived three days in those conditions, let alone a solid week, said Joe. He's paying the Rangers to rescue the two machines and gear, he added, noting the Rangers made the offer.

"Some of the kindest individuals I've ever met."

– 61 –

Storm delays Arctic patrol: Troops won't visit disputed island on 18-day trek

Canadian Press NewsWire, 31 March 2004

Resolute, Nunavut—Diving temperatures and high winds that whipped up 30-metre plumes of snow delayed the departure of the longest one-way Arctic sovereignty patrol in history Wednesday.

But that didn't diminish the enthusiasm of the 20 Canadian Rangers and regular army soldiers for crossing 1,300 kilometres of ice and snow from Resolute to Alert at the top of Ellesmere Island.

"You're filled with a barely containable excitement when you're starting a patrol," said commanding officer Maj. Stewart Gibson.

The party, armed with antique but cold-friendly Lee-Enfield rifles, was to set out on snowmobiles Wednesday morning.

But clear skies degenerated into howling winds and -35 temperatures.

"Sometimes off the top of the hills you can see a 100-foot plume of snow just picked up and held in the air for a while," said Cpl. Doug Stern, a Ranger from Cambridge Bay.

"This is just normal to me," shrugs Stern, who has lived in the central Arctic for 22 years.

"You wait till it's good enough to go, and then you just go."

Gibson's 15 Rangers and five regular army soldiers are charged with showing the Canadian flag in a part of the world that is rarely visited but increasingly contested as climate change opens up possibilities for Arctic shipping and resource development.

One of those flashpoints is Hans Island, an uninhabited pebble in the sea ice midway between Ellesmere Island and the Danish colony of Greenland. The island is the subject of ongoing talks between Canada and Denmark and a recently published photograph shows a Danish flag waving on its rocky shores.

The Canadians would love to plant the Maple Leaf on Hans Island, but their orders say no, said Gibson.

"I would love to be given the order to go over there," he said. "But that's a national decision that's got to be made at the diplomatic level."

Besides, it's too dangerous to try to reach the island over the ice.

"That is such ugly-looking ice, all jumbled up and crunching along," said Stern, who looks out toward Hans Island during the summer when he works at Ellesmere's Quttinirpaaq Park.

"It's not just rough, but it's moving."

This patrol is the latest in a series designed to enhance Canada's presence in the North. It is intended to mark the start of increased efforts to wave the flag over the disputed waters.

Last summer, two navy minesweepers made it as far north as the south tip of Baffin Island, the first naval presence in the Arctic in 13 years. HMCS Summerside and HMCS Goose Bay landed a group of reservists on a small island off the coast.

This summer, the frigate HMCS Montreal is scheduled to sail north, the first large warship to do so in at least a generation. It will carry 200 soldiers and five helicopters for exercises.

Another sovereignty patrol is scheduled next year from Resolute to Prince Patrick Island, on the west side of the Arctic archipelago.

Gibson's party is expected to reach Alert in 18 days. Only seven of the original group will conduct the whole tour in order to minimize impact on the fragile environment of Quttinirpaaq Park.

The party can barely wait, said Gibson.

"Every day, you're saying 'I'm really doing something terribly important,'" he said.

"Then you look at the scenery. It helps the miles go by as you're travelling, the realization that you're making the first footprints in some areas where nobody in recorded history has been before you.

"By the eighth or 10th day, you're getting tired and you're looking forward to a bed that's already made, with running water and a toilet that flushes."

– 62 –

'Brutal' North Hits Canadian Patrol Hard

Adrian Humphreys

National Post, 14 April 2004

The harsh climate and terrain of the far north is taking its toll on the Canadian Forces patrol trying to assert Canada's sovereignty over the Arctic. Two injured soldiers have been airlifted to safety and all but five of the patrol's 16 snowmobiles lie in pieces.

Master Warrant Officer Gerry Westcott, second-in-command of the patrol, was flown out with a damaged leg yesterday after two earlier medical evacuation flights were unable to land because of the weather. Ranger Corporal Doug Stern was rescued Saturday after injuring his back.

"It is brutal punishment on the men and machines," said Master Warrant Officer Westcott. "We were in whiteout conditions and it is hard to tell which way is up and which is down -- it could be a valley next to you or it could be a hill next to you, it's hard to tell.

"We've used up our supply of snowmobiles. We've worn out every ski we have in the north right now. The terrain we go over is probably the roughest you can ever abuse a snowmobile. We are going over boulders the size of milk crates, piles of these boulders with a dusting of snow, to get through some of the narrow passes.

"Every time a machine breaks we are cannibalizing it so we can keep the others going," Master Warrant Officer Westcott said from Eureka, a research station on Ellesmere Island, while on painkillers and waiting for medical attention.

The patrol is en route to Alert, a weather and military installation that claims to be the world's most northerly permanently inhabited settlement, a 1,300-kilometre trek. It is one of several long-range sovereignty patrols planned to bolster a diminishing claim to the remote Arctic territory.

Among the disputes over northern lands are rival claims to Hans Island, a small, barren rock between Ellesmere Island and Greenland, made by Canada and Denmark. Danish warships have been visiting the island and raising their national flag in what Canada claims is its sovereign territory.

Now, about 100 kilometres from Alert, the patrol is expected to push on with the remaining five soldiers.

It could reach Alert by late today or it could take several more days.

Master Warrant Officer Westcott injured his leg when his snowmobile struck an outcropping of rock during a blizzard that reduced visibility to almost nothing, he said.

Warrant Officer Rick Mackwood, a medic travelling with the patrol, was immediately behind him and tended to the damaged knee. The extent of the injury will not be known until he returns to the south at the end of the mission.

"One of the plans was to fly me out to Resolute but, frankly, it only hurts when I move so I would rather them not fly me anywhere to keep the plane up here as a safety factor for the team," Master Warrant Officer Westcott said.

Corp. Stern was injured while the patrol turned around after finding their route blocked.

"Its hard to tell [where you are going] because the contours on the maps don't agree with what is actually on the ground, so we got sucked into a river system... it was absolutely brutal, boulders everywhere," said Master Warrant Officer Westcott.

A front ski on Corp. Stern's snowmobile struck a large rock and he was thrown off and landed on a boulder. He is also in Eureka waiting for medical assistance.

The patrol was expected to cost about $500,000.

It will now likely cost far more.

Rental snowmobiles were needed to add to the army's armada for the first leg of the patrol by 20 soldiers, a mix of regular Canadian Forces soldiers and Canadian Rangers, a military unit comprising mostly aboriginals. Just seven pushed on for the second leg of the mission.

The harsher than expected conditions have meant repeated assistance from a Twin Otter aircraft providing support and supplies for the patrol.

"It is beautiful -- you couldn't find a better postcard than this anywhere in the world, it's the most beautiful scene you'll see -- but pretty rough to go through. It gives a pounding. Most of the guys are pretty worn out," said Master Warrant Officer Westcott. "But it won't stop them. They'll get the mission done. They will reach Alert."

Once in Alert the patrol will fly back to Resolute.

– 63 –

Rangers Assist in Attawapiskat Evacuation

Sgt Peter Moon

The Voice (Timmins), 1 June 2004

Canadian Rangers and a Canadian Forces Hercules aircraft played key roles in the evacuation of 1,160 residents of Attawapiskat when the rising waters of the Attawapiskat River threatened to flood the community during spring break-up.

Ice jams raised water levels on the river by more than 30 feet and water and huge sections of ice started coming over parts of the river bank. The threat of serious flooding and danger to life caused the chief and council to declare a flood emergency.

With only two police officers in the settlement of 1,600, the local Canadian Rangers played the lead role in transporting residents to the airport and getting them onto a shuttle of evacuation planes, including a Hercules from 8 Wing at Canadian Forces Base Trenton.

Chief Theresa Hall was among the evacuees on the first Hercules flight out of the community. As it took off, she said, the passengers all cheered and applauded with relief. The air force airlifted 432 people to safety in four flights. Evacuees were flown to Moosonee, Timmins and Cochrane. Some of the people flown to Moosonee reached Cochrane on a special evacuation train.

About 450 people refused evacuation. The Canadian Rangers set up an operations centre and mounted around-the-clock patrols to monitor water levels and assist with securing the community. They also set up road blocks and organized a Sunday church service for residents who remained in the community. When arson was suspected as the cause of a fire that destroyed an evacuated house, rangers boarded up broken windows in evacuated homes to prevent looting and reduce the risk of more fires.

After five days on constant call, during which most worked 21-hour days, the 13 local rangers were exhausted. "We were zombies, we could hardly think, we were so tired," said Sergeant Sally Louttit, the Attawapiskat patrol leader.

She said the training the rangers had received from the Canadian Forces was invaluable in helping them cope with the emergency.

Warrant Officer Robert Smith and Sergeant James Doherty, instructors with 3rd Canadian Ranger Patrol Group at Canadian Forces Base Borden, flew into Attawapiskat with additional rangers from the nearby communities of Kashechewan and Fort Albany to assist the local rangers. Rangers from Moose Factory assisted emergency officials looking after evacuees lodged in Moosonee. A ranger who was out of Attawapiskat when the emergency took place went to Timmins to assist officials looking after evacuees there.

"The Canadian Rangers worked long hours and helped keep things orderly and peaceful," said Deputy Chief Fred Wesley. "They really helped us. I'd like to say thanks a lot to all of them."

Chief Hall said the response and support for the people of Attawapiskat from all levels of government was impressive.

The emergency ended when the flood waters dropped suddenly and the community was spared serious damage. Rangers met returning evacuees at the airport and arranged transportation for everyone to get back to their homes.

"The response of the Canadian Rangers highlights the relevance of the Canadian Forces and the fundamental need to retain an operational presence in the northern regions of our country," said Major Keith Lawrence, commanding officer of 3rd Canadian Ranger Patrol Group. "The actions of the rangers were of the highest order. We should all take pride in their accomplishments."

– 64 –

Operation Saglek

Jamie Baker

The Labradorian, 20 September 2004

Members of the Canadian Rangers in Labrador teamed with the Canadian Navy earlier this month to complete Operation Saglek.

The operation represents one of two inspections the Ranger Patrol makes each year at northern warning radar sites along the north coast of Labrador.

The Rangers are under contract with NORAD to complete the inspections—one in summer, one in winter.

"In all, we spent 21 days at sea—it was a great trip," said Sgt Bob Toms of the 5th Canadian Ranger Patrol Group. "It's actually, as far as I know, the first time we've coordinated with the Navy in this way.

In winter, we do the checks by snowmobile and in summer we usually hire a boat, get dropped off at the beach and hike to the sites to do a physical inspection. This year, we were very pleased to be able to complete the checks with the Navy aboard the HMCS Goose Bay."

The five northern warning sites in Labrador include Tuchialic Bay near Makkovik; Big Bay, near Hopedale; Saglek; the Kiglapait Mountains south of Okak Bay; and Cartwright.

"The site at Cartwright is actually checked frequently by the Rangers, as it can be accessed by road.

"The sites are old, but they've been updated over the years and we are told they provide excellent coverage. Saglek, for example is a massive facility. Following the 9/11 attacks, the United States actually sent a crew in there to man the facility."

Sgt. Toms says checking the sites can prove to be a daunting task in summer given the challenging terrain along the north coast.

"Inspecting the sites in summer is quite a task. You'd never be able to walk up to them from the beach where we land, so we actually have to hike up to 18 kilometers around the back to get up to some of the sites."

Besides completing Operation Saglek, the Rangers and the crew of the HMCS Goose Bay also provided an invaluable experience for Junior Rangers in north coast communities.

"We stopped at all communities along the north coast where there are Junior Canadian Ranger patrols. We launched the zodiac, brought the Junior Rangers aboard the ship, gave them a tour and they got to meet the Captain and crew. They simply loved it."

The Ranger members aboard the vessel also had the opportunity to be a part of the HMCS Goose Bay's first ever crossing of the Arctic Circle.

"The weather was good, we had lots of supplies, so the Captain made the decision to go north and cross the Arctic Circle. We did tours of Baffin Island and later we went south and met the HMCS Montreal—so really, we crossed the Arctic Circle twice."

"We then went on to Goose Bay for the open house, and once that was over, we were dropped off home and the ship continued on to Halifax."

– 65 –

Canadian Rangers Help British Forces Test New Equipment

Sgt Peter Moon

Wawatay News, 10 March 2005

Two Canadian Rangers from Constance Lake received praise for their on-the-land skills from a group of visiting British forces.

"It's always nice to work with other forces," said Sgt. Heath Blewitt, a British Royal Marine expert in survival and winter warfare, "but to work with the Rangers and see how they work in the woods is a chance few people ever get.

They were fantastic. They are experts; they are unique. They use skills taught them by their fathers and grandfathers and they passed some of that down to us."

He was speaking about Sgt. Albert Sutherland and Master Cpl. Stanley Stephens, Canadian Rangers from Constance Lake First Nation, who worked with the 21-member team from the British forces while they cold-weather tested new communications equipment in the bush north of Geraldton.

The two Rangers showed the British visitors how to cope with the cold, move around in the waist deep snow, how to operate snowmobiles, set snares, build fires and emergency shelters, and other bush craft.

The British team consisted of 17 Royal Marines and four members of the army and Royal Air Force. They spent more than three weeks conducting cold weather trials of the British forces' new $4.4 billion Bowman communications system.

Earlier trials were conducted in Oman, where temperatures soared above 50 C, and the jungle in Brunei, where the equipment faced heat and extreme humidity. Royal Marines normally go indoors or into tents when temperatures drop below 30 C, only staying out in colder temperatures when it is operationally necessary.

Under the watchful eyes of the two Canadian Rangers, the trials team spent a night in emergency shelters when the temperature dipped to 35 C.

"The lads were all very proud next morning," Sgt. Hewitt said. "They'll probably never get the chance to do that again." The two Rangers said the British troops knew little about Canada or Aboriginal Peoples but were eager to learn about their culture, where they lived, the clothes they wore, and what they ate.

The only problem was understanding the distinctive accents and dialects of the British troops. "There's a real language barrier," Sgt. Sutherland said. "Half the time I don't know what they are saying. You have to listen very carefully."

Helping the visitors "feels good, because they really appreciate what you are teaching them," he said. "We're here to make sure they don't freeze." The trials ended with a surprise visit from 10 Rangers from Constance Lake. They provided the visitors with a feast of moose, goose and bannock, a drum song of welcome, and individual Aboriginal headdresses and wall hangings to take home with them.

– 66 –

Wide-open Rangers

The Labradorian, 25 April 2005

Canadian Rangers Patrol Commander Lieutenant Oswald Dyson thought it would never happen.

Churchill Falls to Nain and back on an untested route taking in Smallwood Reservoir and remote stretches of Quebec, a winter adventure to remember.

The idea for the trip began in 1997 when the Canadian Rangers were tasked to deliver a commemorative axe by snowmobile to Nain as part of a celebration of John Cabot's discovery of the New World.

But the attempt, of which Lieut Dyson and Sergeant George Strickland were a part, was eventually abandoned.

"A lack of snow meant the '97 trip had to be aborted and we returned to Churchill Falls nine days after we set out," Lieut Dyson told *The Labradorian*.

The axe was eventually delivered by plane but Lieut Dyson's desire to make the trip just wouldn't go away.

For years he waited for another chance to try the epic route.

"In June, I received a commitment from the Canadian Ranger Commander in Gander to help us with another try at completing the trip," he said.

"This year the timing was our own decision and in conversation with Inuit from Nain it was decided that late March was the right time to go."

On Friday, March 25, Lieut Dyson and his team of rangers Sgt Strickland, Corporal Joe Merrigan, and rangers Boyd Bessey, Greg Anderson and Neil Acreman set out for Nain, filled with anticipation and determined not to fail.

Here's his diary of the journey.

Friday, March 25

"We encountered stormy weather and were caught in slush for a couple of hours. The 100 kilometers across the Smallwood Reservoir was travelled in zero visibility. This certainly gave us an appreciation for GPS. Nevertheless, we covered 180 kms and set up tent on the George River in Quebec the first night. The terrain was similar to the areas around Churchill Falls."

Saturday, March 27

"We travelled about the same distance and stayed in a cabin at Lac Leif, Quebec. The cabin was dilapidated and filled with snow but we made minor repairs and shoveled out the snow. It was comfortable enough.

The terrain was now changing as we were at the upper edge of the tree line."

Sunday, March 27

"Windy with zero visibility at times, we stayed at Lac Leif. Because of the barren land, weather was now critical for us to go any further. Each night I contacted the control room at the Churchill Falls generating station by satellite phone for the weather forecast. On Sunday night we were informed that the forecast predicted fine weather across Labrador the following day."

Monday, March 28

"We discussed the trip as a group and it was decided we would put on a big push the next day to cover the 300 kms to Nain. We could travel no less than 150 kms because there would be no shelter or firewood. As we turned east towards Nain on Lac Brisson, Quebec, the terrain changed drastically. It was basically flat with mounds of gravel and rocks and no vegetation.

"One of the rangers summed it up well when he compared it to the landscape on the moon. Travelling along this section of our journey, I remembered stories of Inuit perishing in this area, of being stormbound for days and of winds so strong it blew over snowmobiles. This is the land God gave to Cain. It certainly looked the part.

"After travelling east approximately 70kms, the terrain changed again, with more hills and valleys. We now began seeing herds of caribou. The closer we got to the coast, the hills became bigger and the valleys deeper.

"The rugged beauty of the land was indeed something to behold. As we neared the coast our travel became somewhat dangerous because we had to travel down several waterfalls.

"However the perfect weather held and we arrived in Nain about 6 pm Monday evening. With a little celebration that night, our thoughts turned to the return trip."

The Trip Home

"We left Nain for home via the coast the following morning to Makkovik, Happy Valley-Goose Bay and then to Churchill Falls. After seven days and over 1600 kms we arrived home Thursday at 6pm.

"This route to Nain had never been completed before. There were so many variables that could have prevented the completion or delayed the trip for days.

"Our conditions were near perfect at the critical stages of the journey.

"I guess we had Lady Luck on our side."

"All of the participants thoroughly enjoyed the experience and took great satisfaction in using their map, compass and GPS skills taught by the Canadian Ranger program.

"I want to thank George, Joe, Boyd, Greg and Neil for making the trip possible."

– 67 –

"The Place at the Edge"

Nathan VanderKlippe

Edmonton Journal, 1 May 2005

Isachsen, Nunavut—A little more than a thousand kilometres from the North Pole, a line of snowmobiles snakes through a blistering wind. Behind each snowmobile, a wooden sled called a kamotiq bounces its load of jerry cans, tents and food across the rough and uneven ice of the frozen Arctic Ocean.

Fluttering from the last snowmobile in line is a Canadian flag, its bright red maple leaf the sole splash of colour in a landscape as bleakly monochromatic as it is majestically vast.

This is the land that silence owns, a place where jagged sea ice runs into sharp gravel cliffs and gently rolling tundra hills, all hidden beneath a windswept coat of snowy sameness that obscures the border between land and sea.

The snowmobiles ride over it all, the flag the sole indicator that this land belongs to someone and that, though the men are closer to the Russian coast than Ottawa, this is Canadian soil.

"We're trying to exercise sovereignty within the North," says Warrant Officer Randy Cox, one of nine regular force soldiers participating in a $1-million exercise that has brought the soldiers and 13 Canadian Rangers to this barren corner of the Arctic on what's called an enhanced sovereignty operation.

"In a nutshell, that means we just want to show a presence up here, in the best interests of Canada," he said. "So that ever in the event that someone else is trying to conquer this land or lay claim to it, then we can at least say that we've occupied certain areas of it."

It's a claim that has drawn criticism from some quarters, from people who dispute that expensive patrols like this one do anything to further Canada's claims to the Arctic and the thousands of islands and waterways in its archipelago.

But this is a critical exercise for the northern military, which is charged with protecting the 40 per cent of the country's land mass that lies north of 60.

The Rangers are Canada's most important contribution to northern sovereignty. Their 1,500 northern members - all military reservists - in 58 communities monitor the land and waters within a 300-kilometre radius of their hometowns, creating a federal presence with a wide reach over the Arctic that is augmented by some 200 annual patrols.

Those parts that the Rangers do not touch are watched by occasional overflights by surveillance aircraft, and the defence department has begun work on Project Polar Epsilon, which will bring regular precision satellite surveillance to the Arctic by 2009.

But there remain great uninhabited swaths of the Arctic, and part of the tasking for the northern forces is to set "footprints in the snow" on exercises like this one, called Operation Kigliqaqvik IV, after the Inuit word for "the place at the edge of known land."

That perfectly describes the place where the Rangers and soldiers are snowmobiling, breaking into two patrols as they leave Isachsen, the site of an abandoned Arctic weather station on Ellef Ringnes Island, for Meighen and Amund Ringnes Islands, smaller chunks of land to the east.

As they travel, a raging windstorm thrusts them on to another edge: the outer limit of human survival.

Once every hour they stop, pulling their snowmobiles together into a cluster against the wind and consulting a GPS unit for a brief location check. The stop is short, however, and the GPS quickly tucked away again inside a parka to preserve its batteries, which have already been sucked of most of their energy by the frigid cold, a -50C windchill that burns white frostbite circles on to the men's cheeks in minutes.

"As soon as you stop, the wind catches up with you and you can turn into a casualty within 15 to 20 minutes if you're not careful," said Cox.

So the men soldier on, pushing against a hail of snow particles hurled by the wind, not intimidated by the blizzard.

"It's how I grew up, being out on the land," said Paul Ikuallaq, a 22-year Inuk Ranger veteran from Gjoa Haven, Nunavut, whose grandfather was Roald Amundsen, the first man to sail through the Northwest Passage. "We're not scared of anything, because we're right there with the warm stuff -- the Coleman stove and the tent."

As the men drive, the wind continues to shriek at gale force. Sometimes it gusts to hurricane strength, the entire sky assaulting the land and the men, its force unbroken by trees or buildings.

And unbroken by land. Ellef Ringnes Island lies at the western edge of the Canadian Arctic archipelago. From here to Russia, the wind blows untamed over ice that never melts. When the island's weather station was still active, the men stationed here sometimes had to crawl on hands and knees between buildings to avoid being toppled by its ferocity.

But lives spent above the 60th parallel have hardened the men on this patrol against the cold. Some of the Inuit leave their faces uncovered as they travel, but their weather-hardened skin is barely touched by the lashing wind, which leaves only a few small "caribou kisses" -- purple frostbite marks.

David Nanook leads one of the patrols. A quiet Inuk man who speaks little English, this land is his home and he is gifted with a remarkable ability to navigate through the limited visibility. Instead of using landmarks, he finds his way by keeping the wind at one side and using the snowdrifts as compasses.

The drifts, some as hard as concrete, form according to the prevailing winds and point in the same direction; to keep on course, Nanook cuts across each one at a similar angle.

By evening, the men have reached their destination, and stop for another GPS check. Nanook has led them for 84 kilometres, and largely by his use of the wind, snow and sun they have stopped 250 metres away from their target. They patrol the coast for a valley, where they set up canvas tents in the searing wind.

The next morning, they hammer the feet of a metal cairn into the frozen gravel, leaving a permanent marker of their presence for anyone who may pass by here.

For some of the men, this is the longest they have ever been away from home, and the farthest north they have ever travelled. Being here as Rangers evinces a chest-swelling pride for many, who feel they are protecting the territory they call "our land" for future generations.

"I wouldn't want a foreign person to own this land because of all the animals around here," said Manasie Kaunak, an Inuk Ranger from Grise Fiord, Canada's northernmost civilian community. "When we went on the patrol, we saw caribou and there were lots of tracks -- I wouldn't want anyone to hunt those besides Inuit people."

The military, too, sees this patrol as an exercise in guarding Canada's frontiers from unknown intruders who might one day claim parts of the High Arctic for the petroleum or mineral riches that could lie hidden in its frozen depths.

"We're up here patrolling the boundaries of Canada to show that it is ours," said Capt. Brian Wiltshire, the deputy commanding officer of the northern Rangers. "If we can't put people here, then other countries are just going to take it."

But experts say that while exercises like this may perk a few ears at foreign embassies, they actually have very little bearing on Canada's Arctic sovereignty.

The Kigliqaqvik patrol is "next to irrelevant," said Franklyn Griffiths, a retired University of Toronto professor who specializes in Arctic matters. "If we really were interested in sovereignty, we'd get ourselves some ice-strengthened naval vessels able to operate in the ice up there," he said. "And that is not coming on."

Unlike Denmark, Canada has no military ships that can operate in ice-choked waters, limiting the country's ability to enforce its sovereignty over the Arctic.

And the Rangers have been dispatched largely on land, which no one but Canada claims.

"They call them sovereignty patrols but nobody is threatening the land mass of the Canadian North," said Rob Huebert, a northern sovereignty expert and the associate director of the University of Calgary's Centre for Military and Strategic Studies. "Our sovereignty disputes are over the waterways and that's really where we have to have sovereignty patrols."

Asserting ownership over those waters now could become increasingly important as climate change melts the Arctic.

In the past 50 years, temperatures in some parts of the Arctic have risen as much as four degrees Celsius. The thickness and extent of sea ice have diminished by about 15 per cent in the last three decades. Scientists expect the future to bring a continued rise in temperatures, which could melt away the permanent ice in the polar ice cap in summertime by the year 2050.

Ice-Free Waterway Will Spur a Flurry of Resource Exploitation

That could bring a host of sovereignty challenges, as temperatures unlock the Arctic's vast network of waterways, clearing the way for a burst of resource activity in the unexplored hinterlands. It could also open the Northwest Passage to transcontinental marine traffic, potentially exposing the delicate Arctic to environmental disaster since the international community does not recognize Canada's claim to the passage as internal waters. If designated as an international waterway, Canada would be unable to set its own environmental rules on shipping through the passage.

Canada faces two other significant Arctic sovereignty disputes: the U.S. does not recognize our claim to parts of the Beaufort Sea, an area rich in petroleum resources, and the state of Alaska has for the past few years invited oil companies to buy parcels of the disputed zone.

In the far eastern Arctic, Denmark and Canada both claim Hans Island, an otherwise insignificant speck of gravel between Ellesmere Island and Greenland, but one that has highlighted Canada's indifferent approach to its actual sovereignty issues in the Arctic.

The past few years have brought increasing lip service to solving Canada's far northern sovereignty dilemmas. Last summer, Prime Minister Paul Martin travelled to Nunavut to underscore the Arctic's growing importance to Canada, which was again emphasized in the October 2004 speech from the throne, which promised a strategy to "protect the northern environment and Canada's sovereignty and security."

But despite a pledge in the April defence policy statement of steps "to preserve our sovereignty, including that of the Arctic," Canada has done little to actually resolve the Arctic's sovereignty challenges.

Over the past two summers, the Danish navy has sent frigates on flag-planting exercises to Hans Island. Canada has not responded in turn. The

Canadian Rangers, frequently held up as the country's foremost northern sovereignty contribution, have only patrolled in areas that are indisputably Canadian territory. They have gone nowhere near Hans Island, or to any of the other disputed zones.

In fact, the closest the Canadian military has come to Hans Island was two years ago in a chartered commercial aircraft, when Stewart Gibson, the commanding officer of the northern Rangers, flew near the island.

He never landed on the island because, he said, "my boss has not told me to go there. Right now, this dispute between the Danes and Canada is at the political level, with (the department of) Foreign Affairs, and they're trying to resolve it at their level."

But at Foreign Affairs, the various disputes have garnered little attention -- none have prompted bilateral negotiations, and department spokesman Reynald Doiron said there are no plans to send a military presence to Hans Island.

"At this time and place there's no particular reason for either party to either negotiate or to bring it to the International Court of Justice, so therefore any presence over there by either side's military forces would not be welcome," he said, despite Denmark's persistence in sending its navy there.

At the same time, social change is beginning to shake Canada's strongest argument for Arctic ownership: the fact that the Inuit have used this land since "time immemorial."

Today in some places, their presence is dwindling, and several communities along the Northwest Passage route have been left as ghost towns, or are near collapse.

No one is left at Shingle Point in the Yukon, while settlements like Bathurst Inlet and Bay Chimo, both on the northern coast of the continent, have dwindling populations now at just over a dozen people. Sachs Harbour, the only community on Banks Island, is dying away: it lost 16 per cent of its population between the 1996 and 2001 censuses, and is now home to only 114 people.

More worrisome still, the hunting traditions that once pushed Inuit to the farthest reaches of the North are losing a battle against the modern wage economy, which brings sustenance from a cubicle and a grocery store rather than from the land. For those who still do hunt, escalating prices have made it difficult to scrape together enough cash to pay for bullets and snowmobile fuel. Ultimately, that has diminished Canada's presence, and its "eyes and ears" across the tundra.

N.W.T. Premier Joe Handley said the remedy is to invest in the communities themselves, and called on the federal government to invest in tourism and municipal subsidies to help keep hamlets like Sachs Harbour viable.

"The best way of ensuring sovereignty and security is to have good strong healthy communities along right across the Arctic," he said. "It wouldn't be very expensive because people who live in those small communities are not looking for a six-digit salary. They like the independence that comes with hunting and living off the land. It could be a good investment to keep that going ... and much cheaper than bringing in military hardware."

Still, both Huebert and Griffiths see value in exercises like Kigliqaqvik, even if not for sovereignty in particular. By most standards, the Ranger program is impressively cheap, costing about $6.5 million per year to run, and Griffiths says the patrols allow Canada to be better keepers of the land – "in the sense of looking after it, we keep it in good order and we manage and see that it is used properly."

Another part of the value is in showing the military's Arctic deficiencies. On this most recent patrol, for example, the winds and heavy ice fog grounded military aircraft for more than a week, delaying the exercise. One Twin Otter scheduled to operate out of Isachsen never made it there because it was equipped with skis that crippled its range.

"The lessons learned in and of themselves are of critical importance," said Huebert. "They drive home to those that pay attention the fact that we are so severely limited in what we can do in the North that hopefully the policymakers make some decent policy decisions on what we can do to shore up that capability.

"They call them sovereignty (patrols) but nobody is threatening the land mass of the Canadian North. Is it really sovereignty? No," he said.

"But it is a very clear enforcement and presence capability, so it does say to people, 'we're up here and we have this capability.'"

– 68 –

Polar Bear School

Adam Day

Legion Magazine, May-June 2005

The soldiers pulled on their camouflage parkas, packed their guns onto the bus and went north to test themselves against the winter. They went seeking a hostile environment where they could practise winter warfare in the kind of high-pressure survival situation that toughens leaders and forges units.

The soldiers weren't alone in this test. For the first time in Canadian Forces history the remarkable natives of the 3rd Canadian Ranger Patrol Group (3CRPG) travelled south to teach their methods of winter survival. By the end of their week in the snow the soldiers of 2 Combat Engineer Regiment (2CER) would prove to be good students of native-style survival, trading their waterproof tents and Hot Rod sausage snacks for snow forts and caribou on a stick.

This was Exercise Polar Bear. Starting in late January and running until the end of February, almost every able-bodied soldier stationed at Canadian Forces Base Petawawa, more than 1,600 in total, would make the trip to Ramore, Ont., 50 kilometres north of Kirkland Lake, to take part in CFB Petawawa's largest exercise of the year. Each soldier spent an average of five days camped in the snow, living off rations and enduring temperatures as low as -40 C. The soldiers practised snowshoeing, ice fishing, shooting and demolitions. But for many soldiers the highlight of the exercise was 3CRPG's seminars on firestarting, trapping, bannock making, meat preparation and other fine points of winter survival.

"Polar Bear was designed to build basic soldiers skills and (give soldiers) the ability to deal with harsh conditions," said Major Geoff Parker, operations officer for 2 Canadian Mechanized Brigade Group (2CMBG). He explained the general purpose of Polar Bear and how it should help prepare soldiers for deployment to places like Afghanistan. "The soldier's ability to cope and deal with harsh environments, whether it is extreme cold or extreme heat, is basically a mental thing. In extreme conditions like this, leadership becomes key and it's a real focus on junior leadership. It's a great opportunity for them to gain really good leadership experience—the kind they can take forward and use in more complicated scenarios like peacekeeping operations."

After the six-hour trip from Petawawa, the soldiers arrive at the site and waste no time. They wriggle into their backpacks, find their assault rifles, strap on their aluminum snowshoes and make the short trek to their campsite. They dig pits in the deep snow and spread out the old circus-style tents that consist of heavy canvas, a tall metal centre pole and dozens of guy-ropes to anchor the sides and provide shape. The tents have no floor, so armfuls of spruce boughs are collected from the forest and placed inside to provide some insulation from the hard snow. While some soldiers fill their tents with several inches of fluffy boughs to ensure a dry and relatively soft week in the snow, others sprinkle an armload or two and call off the effort, trusting the Gortex bivy sacks they sleep in to keep them dry. The soldiers joke that these tents used to be five-man models until budget cuts turned them into 10-man models. And, to be sure, it is a tight fit. But the tents are warm and tough. They are also pretty good hiding spots when a white, silk parachute is draped over the top for winter camouflage.

Since there's no kitchen, the soldiers will live off individual meal packets (IMPs). Three times a day, the Coleman two-burner camping stove is fired up and a stack of entrees is fished from the boxes of rations sitting outside the tent. A typical breakfast includes an entree like ham and scalloped potatoes or hash browns and sausages. Lunch may be penne with beef while dinner might be lasagna or salmon fillet. Regardless of which small silver entree packet you get, the experience is largely the same. Since at least 40 per cent of the IMP's 3,600 calories comes from fat and since much of that fat is concentrated in the entree, each main course is coated in a slick gelatin—variously congealed—that tends to dominate the eating experience.

But foods like this, high calorie and fatty, are a traditional northern staple. When living outside during the winter the body has to work extra hard to stay warm, and it just so happens that fat is the richest source of calories. During Polar Bear, the nights averaged about -20 C and the days hovered below -5 C. Among the native Rangers who came to Ramore to teach survival skills to the Petawawa soldiers, the high-calorie entrees were a superb treat and coveted like candy.

The Rangers came from communities so deep in northern Ontario that no roads go there, places like Peawanuck, Kashechewan and Kitchenuhmaykoosib Inninuwug. The Rangers were hugely popular among the soldiers and it was easy to see why. In everything they did they had a sense of implacable calm and the kind of profound confidence that comes from years of experience. In fact, and this may be hard to believe if you didn't see it, the Rangers had a sort of legendary quality about them. Consider some of their stories: Master Corporal Mathew Gull is a giant of a man who is frequently called on to defend his community of Peawanuck from aggressive and extremely dangerous polar bears. Gull has killed six now, the first when he was 14 and he recently killed one with a small .22 calibre rifle. Ranger Ralph Wayne Mekanak is called Duke because he is uncannily accurate with his Lee-Enfield .303 rifle—numerous eyewitness reports have him repeatedly knocking the gold centre out of a toonie at more than 50 metres. Duke says he never misses and people believe him. Ranger Terrance Meekis, the youngest of the group and known more commonly as Junior, proved to have fire starting skills so advanced they resembled magic.

With a crowd of veteran engineers standing around him, Junior sat down on an upturned log and gathered some twigs into a pile. He took out a magnesium fire starter and said, in a distinctly calm voice, that this was how to start a fire. Just then the wind whipped in hard off the frozen lake. It must have been about -15 C, but Junior didn't flinch. He put his hands around the semi-airborne twigs and, without pausing, struck down some magnesium shavings and then sparked a small glowing fire. In a few moments, despite the wind, he had a good-sized campfire. It was a remarkable feat that took a great depth of unseen knowledge. It was enough of a display to quiet the normally boisterous soldiers, until Sapper Colin Lagaarden exclaimed "that was awesome!"

"I don't think anything could freak them out. And that's what I like about them," said Lagaarden. "I love how they hang onto their heritage so much; they stay so strong with it. They've learned from their fathers and their grandfathers and their great-grandfathers. It's just amazing. The knowledge they have—the survival knowledge—is unbeatable."

The Rangers were formed in 1947 to help defend northern Canada during the early days of the Cold War. They provide a military presence in the North, reporting unusual activity, performing community service and supporting military and rescue operations. Recruited from local natives and Inuit, there are currently 4,500 Rangers in 165 communities spread across the Far North, from Goose Bay in Labrador to Tuktoyaktuk, N.W.T.

"These guys are the leaders of the northern communities," said Captain Guy Ingram, a former military policeman who's now second-in-command of 3CRPG. "The rangers tend to end up being the volunteers for everything. They're the search and rescue experts. They're the guys who get involved with the fire department. They're volunteering to help youth groups and stuff like that."

Despite their leadership role and the tremendous respect they receive from the regular forces soldiers, the Rangers are distinctly unlike any other Canadian Forces unit. They receive only seven days of basic training, have baseball caps as official uniforms and, somewhat distressingly for many CF officers, elect their leaders by voting. So far, 3CRPG leadership has allowed the Rangers to continue their tradition of electing their own patrol leaders, but, like in most modern democracies, they do try to persuade the electorate against rash or uninformed choices.

"If a large family shows up at a meeting they may vote in the wrong person, a person that doesn't possess the right skills to be the leader of the patrol," said Ingram. "However, we don't tell them to vote for any one person, though we do tell them the skills that are required to be a leader and we tell them there are people in the community who have those skills."

The 2CER soldiers spent the better part of a day rotating through the Ranger teaching stations. In addition to fire starting with Junior, they also learned how to build strong and warm shelters out of snow and spruce boughs. Some of the things learned at this station were put to use on the final day of the exercise, when the soldiers went out into the woods in small groups to build shelters and spend the night outside.

At another Ranger teaching station the soldiers learned how to make traditional bannock, a kind of heavy bread that's wrapped around a stick and roasted over a campfire. They also learned how to fillet caribou hind, skewer it on a stick and roast it in the flames; although many of the soldiers chose to eat the caribou raw, in the preferred native style.

During the course of their morning with the Rangers, some longtime members of 2CER had a reunion with Mathew Gull, who they'd last seen during a unit visit to Gull's hometown in 1997. During that visit, 2CER built the community a schoolhouse, which Gull's daughter will be attending in the fall.

The members of 2CER certainly remembered Gull as well, because it was Gull who was instrumental in the rescue of a 2CER soldier who'd become hypothermic during a reconnaissance patrol in the extreme cold. "I noticed one of the privates was limping," recalled Gull. "I went up and asked him if he was cold. A major saw what was happening and said the soldier would be fine." But Gull knew the signs of hypothermia, and knew something had to be done. "I stepped in and took care of him. I gave him all my spare stuff and we got him warm. The major was shocked at how organized we were."

Though the 2CER soldiers certainly appreciate the Rangers, they didn't come all the way to Ramore just to sharpen their survival skills; they were also here to do what the engineers do best, building things and blowing things up. During one exercise, a squad of engineers went out during the day to build small bridges using lumber cut from the surrounding forest. Later, another squad was given the mission to attack and destroy those bridges using C4 plastic explosives.

The mission sounded straightforward, but navigating on a moonless night in waist-deep snow proved to be quite a test. Though one or two errors set some soldiers off course, all the squads eventually found their targets, set their charges and blew up the bridges on schedule.

During another event, on a cold morning in mid-exercise, the soldiers underwent immersion training. Seven volunteers, dressed in bright red survival suits, walked out onto a frozen lake and jumped into the open water. With the men bobbing in the lake, the instructor explained what to do. Venturing out to attempt a rescue is dangerous, and so the first step is to encourage the victims to save themselves. Though getting enough grip on the wet ice to haul your rapidly freezing self out of the water seems pretty unlikely, there is a technique. It's a full-body manoeuvre called the dolphin kick and it actually works. With maximum effort and a lot of splashing, the swimmers get up out of the water and, looking distinctly like a group of seals, roll across the ice to safety. The crowd of watching soldiers erupts into a chorus of seals barks. Everybody laughs.

Laughing and entertaining each other is something the soldiers spend a lot of time doing. Almost every free moment is filled by reciting favourite lines from movies, debating the merits of various actresses and, most of all, telling improbably funny stories about the past. Several of the most experienced soldiers, the sergeants and warrants, seem to have developed material and routines, moving from group to group like professional entertainers. These men are widely admired.

The comedy routines also serve a purpose. On the second day of the exercise, a warrant officer went around the tents telling funny stories about the men who'd gotten sick the night before. Apparently all of them had eaten a particular ration. Shortly after the warrant's visit, a rumour spread among the tents that the tarragon chicken strips are bad. Only the bold and strong-stomached had the tarragon chicken after that.

The system of spreading information through speculation is another favourite pastime for some of the soldiers. While the engineers are out in the forest building the shelters for the final night's survival exercise, the rumour comes around that fresh rations are inbound. And for a few minutes it seems like a sure thing—confirmed by the oldest and wisest among us—that dinner will be hot and fresh and it'll be at 6 p.m. Not only that, but we'll be having a rum ration. Sadly, 6 p.m. comes and the trucks don't show up. Instead, it'll be a bag of saucy fat or a bag of fatty sauce, depending on what you get.

The lack of fresh food, however, did nothing to dim the soldier's spirits. On their final night of the exercise, spent out in the shelters they'd built, most of them stayed up late telling stories around the campfire. This was unit bonding in action, exactly as the planners hoped. And things only got better when out of the total darkness came a yell that the rum ration was here.

In the end, Exercise Polar Bear was a success. No soldiers lost any body parts to frostbite or disappeared into the wilderness and the 3rd Royal Canadian Regt. completed its daring grand finale, a parachute drop at night onto a frozen lake. As for the Rangers, their participation in Polar Bear was deemed so successful that the Land Forces Central Area Command has directed 2CMBG to set up another Ranger training exercise next year. The soldiers, who can sometimes seem pretty tough to impress, will be very happy about that decision. "Some things don't live up to the hype, but the Rangers totally live up to the hype. Their experience, the way they carry themselves and the respect they give the land—they don't demand respect, just meeting them and training with them you automatically feel that right away," said Corporal Cory Coulson, a 2CER soldier from Cambridge, Ont. "I have a lot of respect and admiration for them, just for what they do and how they live."

"If I was attached to a group of them," Lagaarden chimes in, "I'd be like 'whatever you want me to do man, I don't care, I know it'll be OK.'"

Though it's often said an army marches on its stomach, it seems that for these soldiers respect and admiration may be the stronger motivation. Although a bit of rum also helps.

– 69 –

Northern Patrol

Joseph Hall

Toronto Star, 18 June 2005

The native reservists in the Canadian Rangers use their army training to help others in the impoverished native communities of Northern Ontario, Briefly splitting the Albany River on its northeast rush to James Bay, 160 kilometres downstream, this eye-shaped island was once witness to a massacre. Its name means ghost in Cree. And the bones of the southern Mohawk invaders who were ambushed here by northland Cree warriors some three centuries ago lay for months along the island's sandy shore and along the banks of the Cheepay River, which empties into the Albany nearby. Their spirits are said to haunt the place still.

"This is the story that was brought down to us," says Joseph Sutherland, an elder from the Cree community of Fort Albany, near the river's brackish mouth.

For the Cree, Ojibwa and Oji-Cree peoples of Ontario's far north, the ghosts of the past are the least of their worries. Today, Indians who inhabit more than 50 isolated communities in Ontario's north are being haunted by modern demons. There's alcoholism, poverty, gasoline-sniffing and child abandonment. There's ignorance, violence, rape, murder and epidemic diabetes. And yet there's a set of modern warriors, almost unknown in the big cities to the south, who are helping bring hope to the people north of the 50th parallel. Armed with World War II vintage, bolt-action rifles and uniformed in dirty, red sweatshirts, caps and toques, it's a ragtag band of soldiers to be sure. But the 3rd Canadian Ranger Patrol Group - a band of some 400 First Nations reservists who represent Canada's armed forces in Ontario's far north - has become a scarlet glimmer of hope in an often-blighted human landscape. And over a recent four-day period, some 57 of them gathered on Cheepay Island in the largest exercise ever of this military outfit - surely the most unusual in the Canadian Armed Forces. Under a lowering sky that promises snow but delivers nagging drizzles of rain for much of their frigid stay, the Rangers, both men and women, will take part in a series of competitive drills that, for all their meticulous planning, have a madcap air.

From the 13 community-based Ranger patrols represented on Cheepay, the soldiers will be split into five teams that will adopt names like Shania (for the songstress Twain), River Rats and Wolverines. They'll compete in target-shooting, using their standard-issue No. 4 Lee Enfield .303s. There'll be skeet-shooting, first-aid, map-reading, compass navigation and the art of knots and pulleys. But the bulk of the drills focus on the hunting, fishing and survival skills that the Indians themselves bring to the Rangers table.

They will build goose blinds - and fashion instant flocks of wooden geese decoys to go with them. They'll compete in geese-calling contests, with some gratuitous moose calls thrown in. They'll fish - how they'll fish! - build lean-tos and conjure roaring fires in seconds. They'll prepare bannock - a bread-like concoction introduced by Scottish trappers during the Hudson's Bay Company days - over open flames.

And they'll try - in what is likely one of the most important forums left for preserving the culture of Northern Ontario's Indians - to stem the steady erosion of these traditional skills.

Major Keith Lawrence is the Toronto-raised commanding officer of the 3rd Canadian Ranger Patrol Group.

The Cheepay exercise cost $150,000 to stage and took a full year to plan, says Lawrence, 40, who is white and whose past postings have included stays in Cyprus, Syria, Jordan, Congo, Uganda and Israel.

Outside of its small and largely white regular army leadership, which is stationed at Canadian Forces Base Borden near Barrie, the 3rd Ranger Patrol Group is almost 98 per cent Indian.

Located in 15 different northern communities - two of which could not make the Cheepay gathering - each patrol consists of between 17 and 36 Rangers who receive 10 days of basic army training each.

This instruction, most of it provided in their own communities, includes rifle training, general military knowledge, map, compass and GPS navigation, first aid, search and rescue and communications. The reservists may also have training in flood and fire-evacuation planning, major air disaster assistance and rifle firing exercises.

Rangers are paid between $78 and $136 a day for performing official duties, which do not include policing. They receive compensation for the use of personal snowmobiles and other equipment.

Each community patrol has a cargo bin containing first-aid supplies, a short-wave radio and GPS locator. The patrol must meet at least six times a year. Beyond that, training or operations are more or less ad hoc and largely voluntary.

Despite any training and gear they might receive, however, the Rangers bring a cultural perspective to the organization that stubbornly defies traditional military discipline.

From the window of the 1964 de Havilland Beaver float plane, Cheepay Island seems to appear out of nowhere.

An hour's flight north of Hearst - some 1,135 kilometres north of Toronto - the island sits just above a vast stretch of muskeg; swampland that will unleash billions of mosquitoes into the air in the coming weeks.

Huge chunks of ice from the Albany's spring breakup still line the island's shores during the Ranger's mid-May stay, and temperatures, which mostly hovered around zero, would sometimes drop to minus-10C.

The river runs up against a wide, weedy, rock-strewn beach, where the Rangers parked the 25 motorized freighter canoes that brought them here from staging areas at Constance Lake to the south and Fort Albany to the northeast.

Up a set of dirt stairs, carved into a steep hill at the beach's edge, amid tall cedars and bare poplars, the Rangers have pitched their camp.

About a dozen large, white canvas tents have been slung over long, softwood poles, which were carved out of the forest that covers most of the kilometre-long island.

The tents housing the 11 regular army headquarters staff who lead the Rangers are supplied with canvas cots and military-issue, cold-weather sleeping bags.

But lodgings for the 57 native Rangers (and two Junior Canadian Rangers attending Cheepay) are furnished with the blankets, tarps, foam mats and improvised wood stoves that they'd use during their regular trips into the bush.

Large tarpaulins, strung from trees and poles, protect the half-dozen fire pits that will burn throughout the Rangers' stay. And the sound of wood-chopping will provide a persistent background tempo to the proceedings.

Food on the island consists mostly of the Army's vacuum-sealed individual meal packs of gourmet delights such as chicken teriyaki, cheese tortellini and ham omelette.

But moose, caribou, goose, bannock and, of course, fresh fish are on many menus around the Ranger campsite. Cigarettes dangle from the vast majority of mouths.

In an outfit where fishing can fulfill military commitment and where sergeants are elected by their men and women, you'd hardly expect to see precision drilling.

And on Cheepay there's none. There's also little saluting - except in jest. There's no marching, no shiny brass fastenings or, very often, any buttons at all.

The bugle-boy sense of urgency is definitely absent, as are rules regarding hair length, grooming, cleanliness and "snap-to-it," command-chain respect. Even a straight muster line seems beyond the care or capacity of some Cheepay participants.

Present in abundance among the Rangers, however, are the traditional skills bred by centuries in a grudging, harsh and killing land.

And these skills, except for the Ranger intervention, might well be on the way to extinction in the troubled Indian communities of the north.

They're troubles that Robert Gillies knows well.

Gillies, 37, tells bad jokes - constantly. They're groaners and they'll typically cast hapless Mohawks as the dupes.

But his humour is partly a defence mechanism, jokes and one-liners to hold back recollections from 10 years of policing.

As a former sergeant in the Nishnawbe-Aski Police Service, which patrols most native communities up here, the Cree Ranger from Fort Albany has seen the worst of native desperation up close.

"It's a (way of) coping. I put that (bad) stuff away in my mind where I can't retrieve it," says Gillies, 37, who quit the force in 2003.

"If I kept those things in a (conscious) part of my brain, I don't think I'd be able to function."

Over his policing career, during which he rose to criminal investigator in the service's Cochrane divisional headquarters, he saw the worst of the northern aboriginal calamity: suicide, rape, child abuse, even murder.

In one unwanted memory of his policing days, he recalls a tiny pair of siblings cast off by their parents for the bottle.

"It involved two boys, one of them was basically an infant and the older brother was 3 or 4 years old," Gillies says.

"We, as police, brought the two kids to the hospital ... and you could tell the small child was already looking after the infant. The 3-year-old was looking after the infant." After 10 years as a cop, he says, "I basically had enough."

Gillies joined the Rangers a decade ago and sees the outfit as a major force for good. Indeed, he says, Rangers can often take over community responsibilities in the absence of official band alternatives.

"For example, every year there's usually a flood co-ordinator in the community for (ice) breakup. Every year it's exciting," Gillies says.

"And one time, the flood control committee broke down ... I think the co-ordinator had been drinking. And the Canadian Rangers took over."

The Rangers are lightly trained and are not required to serve overseas in times of war. But they represent a legitimate branch of the armed forces.

Founded in 1947, they were established as a Cold War means to patrol remote northern locales, largely for signs of Soviet intrusions. The force has grown to about 4,500 reservists and is expected to increase to 4,800 by 2008.

They are divided into five different patrol groups across the country, in every province and territory except New Brunswick, Prince Edward Island and Nova Scotia.

Their primary mission remains the patrol and security of the country's northern and coastal regions.

In this time of global warming, with ice-free Arctic waters opening up the possibility of year-round shipping and mineral and oil extraction in the far north, sovereignty over lands now claimed by Canada may come into dispute.

What the Rangers can represent in the Arctic is federally funded feet on potentially contentious ground, Lawrence says.

"You have to show your neighbours and show the world that, yes, you want it and you're actually doing something," says Lawrence, whose Ontario group is the newest and smallest of Canada's five Ranger contingents.

Of course, questions of sovereignty are unlikely to arise in Ontario.

But Rangers here can still have a role in the issue, says Lawrence, who took command of the province's group last year.

"While we don't have a pure sovereignty task here, what we do have is the ability to train soldiers to go north," he says.

"It's very expensive to send soldiers into the Arctic. And you can get very similar weather conditions here. All the same (training) principles apply, whether it's navigation or how to survive on the land."

Perhaps more than any other branch of the Canadian armed services, however, Rangers bring their military status and training directly to bear on their own communities.

By dint of their army instruction and, in many instances, an elevated sense of social responsibility, Rangers often become leaders in their communities, Lawrence says.

They commonly provide organized help when flooding or forest fires threaten northern communities, he says. Many volunteer to serve on local fire departments and join in band councils. And most have steady jobs in communities where full-time employment is by no means the norm.

"They're the doers in their communities," Lawrence says.

As such, the Rangers have become something of a social program, Lawrence says - a federal means to send expertise, leadership and extra cash into impoverished and often chaotic communities.

There's also a Junior Rangers element to the outfit, a type of cadet service that, among other things, helps steer native kids away from the delinquent activities that are commonplace in northern communities.

Another role the Rangers perform in Northern Ontario - one that they're enormously adept at - is search and rescue operations.

Rangers-led rescue missions into the bush or out onto the rivers, often conducted in the most brutal weather conditions imaginable, may number in the hundreds each year.

But with most Rangers loath to fill out paperwork describing their actions, or reluctant to embarrass those who got lost, most of their rescue stories remain untold, Lawrence says.

Last year, Rangers also played the key role in an evacuation of Attawapiskat near James Bay, where the threat of flooding forced 1,154 people from their homes.

An eight-year Ranger veteran, Vicky Edwards is a native from the community of Fort Albany. But her eyes belong to the Scottish highlands.

Edwards' great-grandfather was a Scottish fur trader whose genes have come back with a vengeance in the 27-year-old Cree. Her light hair, freckles and hazel eyes all speak more of thistles than muskeg.

But when Edwards fishes, she's all Indian.

Like most of her Ranger colleagues, Edwards will fish at the drop of a hat. And like many of the Cheepay cohort, she does it like a machine.

Having travelled about 10 kilometres up the Albany for an evening of angling, Edwards and two Ranger colleagues turn their 22-foot freighter canoe into a tributary stream and ease it in by a beaver lodge close to shore.

Then, the frenzied casting begins.

With lines whizzing across each other, the Rangers plunk their lures again and again, with amazing precision, into the reedy waters surrounding the beaver house.

"Fish on!" Edwards yells at least 10 times during a half-hour of fishing.

In that time, she lands five good-sized pickerels and one large pike, which will flop around on the floor of the boat all the way back to camp.

As darkness descends, the Rangers turn for home, a 30-horsepower outboard powering their trip.

Told in no uncertain terms to be back by 9 p.m., they're already running late.

But when they spot an American eagle circling a stretch of shoreline, they pull the boat up and break out the poles again.

"That means there's probably fish here," Edwards says.

And if it's a contest between fishing and following orders, the fishing will easily prevail.

– 70 –

Hudson Bay Quest

WO Wade Jones

Western Sentinel, 11 August 2005

It started off as just a dot about 8 kilometers away on the featureless sea of snow called the tundra, but it was the first of the dog mushers to arrive at Nunala, Man.

This was the half-way point and the third check point of the second running of the annual Hudson Bay Quest (HBQ). It is an extremely challenging 400 km dog sled race from Arviat, Nunavut to Churchill, Man. along the frozen coastline of the Hudson Bay. The HBQ commemorates the historic trading route between Churchill and Arviat. Mail and supplies were carried by dog sled between the two communities until the 1950's.

The planning for Ranger support to the HBQ started in December with a request from race organizer Dave Daley to the Manitoba Detachment Commander, Master Warrant Officer Doug Colton. He requested that the Churchill Ranger Patrol provide support to the race route from the 60th parallel to Churchill at designated check points as well as pick up and escort the race veterinarian and assistant from Arviat, Nunavut to Churchill, Manitoba.

This was the start of a long but easily overcome process for the Rangers from Churchill to participate in the HBQ. It started with the usual phone calls and ended with a provision of service contract being prepared to cover all the legal aspects of the Canadian Forces providing a service to a community event like this.

Next was the coordination between 4 Canadian Ranger Patrol Group and 1CPRG, as both groups would provide the same support on their respective sides of the 60th parallel. This was easily accomplished due to the cooperation of the two groups and the common goal of having the Rangers support community events such as the HBQ.

The preparations for the Churchill Rangers and the race started with the arrival of the Det. Comd. MWO Doug Colton and me, Ranger Instructor Warrant Officer Wade Jones, in Churchill March 29. We were to participate in the exercise as observers and allow Patrol Commander Camille Hamilton to run his patrol.

We would only step in if a safety issue arose. The [Patrol Commander] soon figured out that he had two extra workers and put us both to work helping him prepare the Bombardier, which was to be our mode of transport for the next eight days and some 800-900 km.

Now this was the first time in a while that I greased road wheels and checked fluid levels, sorted through spare parts, and loaded the komatik with

90 gallons of fuel plus all of our gear to live on the land for eight days. It brought back fond memories of my days as an APC driver. (Not!)

The [Patrol Commander] had issued his orders prior to our arrival and had his final patrol briefing March 31 to ensure that each section was ready to deploy and mark the route to their respective checkpoints from Churchill to Nunala. Sections would be deployed according to their location and on different days.

Bright and early April 1 our brilliant yellow Bombardier left Churchill accompanied by Ranger Kevin Burke and Trent Watts both travelling by snowmobile. They would man the checkpoint at Nunala and mark the route south to Hubbard Point.

Onto the sea ice we went crossing Button Bay about six km from shore. The ice was a little rough at the start with all the hummocks and pressure ridges but soon it turned flat as a backyard ice rink. Once across Button Bay it was on to Hubbard Point where we left the bay and headed inland onto the featureless tundra.

We made excellent time and arrived at Nunala in about 10 hours. We had left the tree line many hours before and were met with a vast open featureless landscape that played tricks with our senses of direction. The GPS was a very useful tool in maintaining direction and keeping track of our position.

Once in awhile we would see a tree in what seemed the far distance. We would use it as reference point only to realize that it was a small bush only a few hundred meters away. Other times we would do the same to find out it was a stunted tree but took hours to reach.

Our first day on the land finished with the digging out of the small building at Nunala. Nunala means small land surrounded by water in Inuktitut. It was one of the last trading posts owned and operated by the Hudson Bay Company and was in operation from about 1930 to 1950.

It is mentioned in many books that are written about the Bay and travel up and down the coast. It is believed by the Inuk to be haunted and many stories were told of the singing of hymns and laughter heard while sleeping near the building in days past. I can only attest to a lot of really loud snoring, some of which I may have contributed to myself.

Next morning, leaving Rgrs Burke and Watts at Nunala to carry on with their tasks, we headed north again for the 180 km trip to Arviat. Once again across the tundra we travelled but further inland than the previous day.

Caribou were starting to show up and were amazingly tame. We could sometimes see them from 5 km away. They were normally in small family groups of about 30-40 head. Of course we were forced to stop for pictures, as they stood motionless wondering what the heck we were.

Between Nunala and Arviat there is not much to the untrained eye of a southern *kabluna* (white guy) but the locals have names for every little change in the land. To the [Inuit] there are plenty of navigational aids.

If there are no aids then one is made. Inuksuit [plural of Inukshuk] are easily spotted and mark the ancient travel routes that are followed during hunting trips. Inuksuit are made from rocks and the word Inukshuk is an Inuit word meaning "in the image of man" because of the lifelike figures created. It doesn't take long before you realize that the speck on the horizon is a marker and it is at least eight km away. It may be only a couple of feet high but it will stand out like a sore thumb.

Travelling was going well. The only real reason to stop was to fuel up and, well, you know the other reason. The [Patrol Commander] had mentioned several times that the last time he had travelled to Arviat he hit a whiteout only a few miles out of town and had a real hard time finding his way.

We had been travelling for about nine hours, and we could see Arviat, a small hamlet on the coast of the Bay. We were still about 10 km away and needed to fuel up again, so a break was in order.

In the time it took us to dump 10 gals of gas into the tanks, a bank of whiteness descended upon us and we could not see more than 100 meters in any direction. The Patrol Commander was right - we got our welcome whiteout. Once again we navigated 100% by GPS and made our way into town. There we were met by our host for the night and were soon settled into our rooms.

It only took about an hour and Sergeant Jerry Vanhantsaeme from 1 CPRG came to do some coordinating. He was accompanying the 10 Rangers from Arviat for their support to the HBQ.

That evening was also the HBQ start banquet. It was held at the Mar Halluck Hall and it was a real community event. There must have been 500 men, women and children-and I mean lots of children!

There was a community feast and jigging to a local band to mark the start of the race. People were very friendly and were full of questions about our uniforms and who the new *kablunas* were.

Over the next few hours there were introductions, speeches and the drawing of the race start positions as well as a feast of caribou soup, muk-tuk (beluga whale blubber) and bannock. The [Patrol Commander] and I settled for a coffee but the detachment commander was dragged to the large soup caldron for a sample. I leave that tale for him to tell.

The manning of the Arviat checkpoints were covered off at 75, 101, 180, 260, 345 and 395 km, the last checkpoint being the finish line.

Next day was race day. We had picked up Dr. Evan Fisk of Winnipeg and his assistant Kelsey Eliason of Churchill.

We visited each checkpoint and talked with the Rangers from Arviat who had deployed the previous day to ensure the route was well marked. We also met several other small groups of people along the trail who had been out wolf or caribou hunting for the past few weeks.

As vast and as empty as the land seemed, it is well travelled by snowmobile and Bombardier either hunting or hauling cargo between communities.

Cargo is not limited to supplies for living. If you want a new vehicle, there are plenty of people who will pick it up at the end of the railway in Churchill. They will put it on a huge komatik and tow it across the tundra and frozen sea to places north like Arviat, Rankin, Whale Cove or Chesterfield Inlet for a nominal fee.

It is a little amusing to see a new full-size Dodge cargo van on a sled.

Once again navigation was tested in a whiteout for the later part of the day. We finally broke the storm about 10km from Nunala as we neared the coast. Rgrs Burke and Watts had successfully marked the trail south to Hubbard Point but had been soaked to the bone by a freak rainstorm the day before, another example of the dangers of travelling in the North.

Both Rangers had been very busy and had completely set up the small building to be a home away from home complete with portable propane furnace. We had spent about two hours two days earlier digging out the snow filled post. This included polar bear tracks in a corner of the building where the bear had obviously found a place to get some rest recently.

Rangers from Arviat had also arrived that morning and had set up their tent in the lee side of the building. A few hours later the remainder of the Arviat Rangers arrived and set up their tent as well. Communications had already been established using the para comm radios. The Checkpoint 4 Rangers had deployed as scheduled the day before and had encountered no problems enroute or marking the trail to Checkpoint 5, North Knife River.

The last of the Churchill Rangers would deploy the next day to the final checkpoint.

That evening the two [patrol commanders] talked with their respective checkpoints and news soon came from the north that the mushers were having a difficult time in the storm. The two [patrol commanders] continued to relay information north and south over the next 18 hours or so. Communications played a very important role for both patrols. The para comm radios proved to work well but at times the sat phones had to be used as backup.

By the afternoon of day two not all the mushers had arrived at the first checkpoint. During the first 24 hours two had pulled out due to injuries and one musher was missing. [Patrol Commander] Issakiark dispatched Rangers from checkpoint one to backtrack and see in he could be located, and he was. He had become disoriented and ended up close to the floe edge on the bay. He had several sick dogs and was convinced to head back to Arviat.

Word had been sent that morning that the first musher had left Checkpoint 2 and should be expected late that afternoon at the half way point, since the weather had cleared and snow conditions were good.

It started off as just a speck and about an hour later Andrew Panigoniak and his eight dogs slid silently into the halfway point of the race. Over the next several hours Nunala grew to a population of 26 people and 40 dogs.

You could already see by the action of the mushers that they cared for this style of Inuit sledding. Their dogs were their first priority, then the komatik and then themselves. Nunala had become a buzz of activity.

Rangers from both groups were responsible to check each musher for the mandatory equipment they must carry.

The race vet Evan Frisk was busy checking the dogs as they arrived. It was 'dog lover heaven' and I was strongly encouraging MWO Doug Colton to pick out one more to add to the three he already has at home.

He declined on the grounds of probable divorce.

[Patrol commanders] were sending radio messages of who had arrived. All the while the dogs were yelping for their supper. It was quite a show. Soon the flow of activity slowed and everyone was settling into a night routine. Some slept in tents, some in the Bombardiers they had arrived in and one musher built himself an igloo. Noise soon gave way to silence and all had a good night's rest.

Next morning all that could be heard was the sound of barking dogs that were full of energy to get back on the trail. It is amazing. You can tell they love to run as much as the rest of us love to have a good coffee.

They can run all day at about eight miles an hour. When they stop they roll in the snow to get the excess moisture out of their fur then have a bite to eat and drink then curl up in a ball and sleep. They are then ready to do it again.

As the last of the dog teams departed we made preparations to break camp and start the move south to each checkpoint.

Early the next morning we were off again in the yellow Bombardier. I thought for sure that we would catch up with the last musher on the trail but we didn't.

He was at Checkpoint 4, Hubbard Point when we arrived. We were met by second-in-command Dave Lundie who informed [Patrol Commander] Hamilton that all the mushers had passed through and the last one had travelled through the night to get there. He was resting his dogs for several hours there.

Rgr Kelly Turcotte informed the race vet that some dogs had been left by the mushers for various reasons and one had a nasty bite from fighting. So into the cabin went the dog and Rgr Turcott held the dog while minor surgery was performed.

Section Commander Janice Schultz served up a coffee and we had a bite to eat. Things had gone well at the checkpoint with the mushers arriving spaced apart, which allowed the Rangers time to carry out the checking and so on with ease. After loading the two dogs into the dog box and then onto the komatik, we were off for the next checkpoint at North Knife River.

It had turned out to be a beautiful day. We were now back out on the sea ice and, as before, it was very rough going till we were out beyond the pressure ridges and hummocks.

Standing up in the open hatch of the Bombardier I could see something strange moving. It was a long ways off and it was some time before we realized it was a tractor pulling a large sled loaded with building material.

We stopped and talked for a while and found out it was a new tractor being delivered (driven) to Arviat. This was about 300 km north of us now, and would be a long trip at a top speed of seven [kilometres per hour].

It was safe for the mushers but the bomber, as it had become known as, was a little too heavy. So we left the mushers' trail and made our way around only to have to backtrack slightly and cross over the pressure ridges once again before arriving at Checkpoint 5, North Knife River.

Rgr George Lundie met us at the checkpoint where a tent had been set up on the edge of the mud flats, which was covered in a metre of snow. This checkpoint had had a very late night the night before.

Some of the mushers had arrived just before last light out but one was late and Rgr Stanley Spence with his keen eye had seen a light a couple of km out on the bay.

Knowing that locals don't travel at night often, he concluded it must be a musher and off he went to investigate. Sure enough it was. The musher had gotten turned around and had missed the checkpoint and route markers. Rgr Spence gave him some pointers and guided him back to the checkpoint.

It had been a busy late night and an early morning as well, but all were in good spirits. Some of the mushers were up well before first light and pushing on for the last leg of the race.

After about an hour the [patrol commander] gave the word to load up and we were off on the last leg of the trip.

We were soon back on land and the final few km across the Churchill River. At the finish line I looked back to proudly see the Canadian Ranger banner. Nobody had asked for it to be there but I guess the Race Marshall Bruce Andrews was so impressed and thankful for the support of the Rangers, it was his way of saying thanks.

The final musher crossed the line that evening and the remaining Rangers redeployed the following morning and were back in Churchill as a group by 4 p.m. Friday, April 8.

The HBQ was not over yet, there was still the awards banquet to be held Saturday evening at the local branch of the Royal Canadian Legion.

Many awards were given out, but you had to be there to hear how often both communities thanked the Canadian Rangers for their support. From the first place finisher, John Stetson of Duluth, to the most determined musher, Moses Kigutnak of Arviat, Nunavut, a hardy thanks were given to all the Rangers.

The two patrols had been on the trail for eight days, marked and covered more than 400 km of trail, travelled probably 2,000 km in total and ensured that all the mushers and their dogs got safely to Churchill. This was not a small task.

It was accomplished through good planning, co-ordination, leadership, resourcefulness, and above all, the ability to have fun as a Ranger. [Patrol Commander] Camille Hamilton and his Patrol did a first class job and should be proud of their abilities.

– 71 –

Exercise Hudson Sentinel 05: Churchill Patrol

WO Wade Jones

The Ranger, Fall 2005

It has been almost 30 years since a Canadian Forces warship has entered the Hudson Bay and the Port of Churchill, Manitoba, but that's what happened on 17 Aug 05. HMCS Glace Bay and HMCS Shawinigan tied up to the massive wheat terminal dock to a large crowd of curious onlookers. That evening the crew of both ships turned out in their dress uniforms commonly called "salt and peppers" to be given, by Churchill's, Mayor Michael Spence, the highest honour afforded to a military unit: the "Freedom of the City" and a plaque with the key to the town. Mayor Spence declared that 19 August 2005 would be "Canadian Forces Day" in the town of Churchill. On hand for the occasion was Member of Parliament, The Honourable Reg Alcock, President of the Treasury Board, Commodore Bob Blakely, Commander of Canada's Naval Reserve, the Churchill Ranger Patrol, citizens of Churchill, international tourists and the national press. Afterwards the ship's Commanding Officers, LCdr Scott Healley, HMCS Glace Bay and LCdr Etienne Landry, HMCS Shawinigan hosted invited guests, including the Rangers, on board for a meet and greet. Many of the Rangers were very impressed with the crew's hospitality and the friendly atmosphere that made them feel like true members of the Canadian Forces. The Rangers were equally as busy answering the crew's questions about the Ranger programme as the sailors were answering questions about the ships.

The impressive Kingston Class Maritime Coastal Defence ships are part of Exercise Hudson Sentinel 05, a multi-element Canadian Arctic sovereignty exercise that involved the warships and their Naval Reserve crews, an Aurora maritime patrol aircraft, a Twin Otter and personnel from Maritime Command, Canadian Forces Northern Area, 440 Squadron, 405 Squadron, Fifth Maritime Operations Group, Canadian Rangers from 1, 2, 3 and 4 CRPG and the Royal Canadian Mounted Police as well as many civilian newspaper reporters from Canada and other countries.

On the afternoon of 18 August the ships were ready to "turn the screw" as they say and invited many of the town's people including Section Commander Janice Schultz, Rangers George Lundie, Stanley Spence, Conrad Bjorkland, Joan Brauner, and Michael Witmore to experience life at sea for a few hours. The ships crews went through several drills including man over board, turns and so on, showcasing and obviously proud of their fine vessels. The Rangers were truly impressed with the professionalism and the ships' ability to turn so fast. Ranger Whitmore was more impressed at their ability to dock, as he was a little green around the gills for most of the trip. Once back on solid ground, he too was proudly displaying the ships ball cap as a souvenir of his experience.

Included on 19 August in Churchill, was a Memorial service held at the Churchill Branch #227 of The Royal Canadian Legion honouring the year of the Veteran and the Canadian Forces visit to the town of Churchill. Approximately 200 Churchill residents and Canadian Forces members including an honour guard from the ships company attended this service. It was a patriotic service that reminded us all of the sacrifices that our Veterans and serving members have made to make our world and country a better place. An open house and displays of the Aurora and Twin Otter aircraft, the Canadian Rangers and Junior Rangers equipment and guided tours of both ships followed the service. The back drop for the displays at the port that afternoon were thousands of white beluga whales noisily blowing air form the top of their heads as they surfaced sometimes within 30 feet of the dock and ships. The afternoon proved to be not only a chance for the community and ship's crews to see the Rangers but also served as an excellent recruiting venue. Several people showed interest in joining the Rangers programme. During the afternoon of the 20th a friendly ball game was played between the ship's crews and the town. Winning was measured in the friends made and the fun had by all.

The involvement of 4 CRPG in Ex Hudson Sentinel 05 started many months earlier in April 05, with an invitation coming from the Fifth Maritime Operations Group, Halifax to have some sort of participation of the Canadian Rangers in the Exercise. The Detachment Commander, Capt Doug Colton, knew this would be a great opportunity for the Det so once the CO gave his approval he did his thing and pursued to have the Det involved. It (the plan) evolved and when the final OP Order was announced 1, 2, 3 and 4 CRPG were invited to have Rangers participate. Manitoba Det gave the warning order to Patrol Commander (PC) Camille Hamilton of the Churchill Patrol, in early June, giving him the general idea of the exercise as known. He was very happy to hear the Churchill Patrol was allocated five spots, one Ranger Instructor WO Wade Jones, two Rangers, PC Camille Hamilton and Ranger Kelly Turcotte as well as two Junior Rangers Stephan Mudrak and Justin Preteau to sail north. Junior Canadian Rangers Officer (JCR O) Capt Sally Purdon and JCR Instructor WO Jeff Belisle were just as excited as the JCRs themselves to have the JCRs involved.

The five, along with three Rangers and one Ranger Instructor from 3 CRPG, one Ranger Instructor and the Group GSM from 1 CRPG were divided between the two ships; they boarded on 21 August and sailed north.

The hamlets and Ranger Patrols on the west side of the Hudson Bay that were visited during the week long exercise were; Churchill, Arviat, Whale Cove, Rankin Inlet, Chesterfield Inlet, Coral Harbour, cross the arctic circle, Cape Dorset, Kimmirut then the ships moved south to St John's, Newfoundland, then to their home port Halifax, NS. The Churchill Rangers travelled on the HMCS Glace Bay from Churchill to Whale Cove, Nunavut approximately 600 km, where they put ashore in rough seas aboard the RIB [rigid inflatable boat]. A Canadian Forces CP 138 Twin Otter met them and flew to Arviat, Nunavut.

There they spent the night. In the hamlet they were greeted and welcomed by everyone they met and of course asked lots of questions. Next morning they returned to Churchill by air.

From Arviat and points north, when weather permitted, more Rangers' families and friends were taken on board at each hamlet for the day. They were given tours of the ships and put ashore before moving on the next community. The ships sailed far into Canada's Arctic prepared to conduct a [search and rescue] near Cape Dorset with 1 CRPG before they returned to St John's, about one week later.

The scale deployed, presence and importance of the Canadian Forces being in the Arctic cannot be overstressed. Not only did it offer the small isolated communities and Rangers a chance to see the warships and aircraft, it also showed our sovereignty in the north, the main objective of the exercise.

The Manitoba Det was very fortunate to have the Rangers and Junior Rangers spend a day sailing north and fly in a Twin Otter. Both Junior Rangers Stephan Mudrak and Justin Preteau had the time of their lives. Both boys assured me that they were more than capable of safely steering the ship to Whale Cove if the Captain needed a rest. JCR Justin Preteau was wide-awake till 3 AM assisting the Officer of the Watch with handling any problems on the bridge, he was answering the ship's Coxin MWO O'Quinn at 0700 hrs as to why he was still in the rack. Actually he was given an extra 15 minutes just to get the sea salt out of his eyes. JCR Stephan Mudrak was all smiles when he was presented a Bosun's whistle by the Bosun's mate while on the sweep deck. He could be heard for most of the morning practicing the ship's piping. Hope nobody went to action or man over board stations. [Patrol Commander] Hamilton covered every inch of the ship's controls, radars, GPS systems and wanted all the details on power, propulsion, horse power, kilowatts, jewels, AC, DC conversions or anything else he could add to his inquisitive mechanical mind, no sense sleeping if you can learn something. Ranger Kelly Turcotte was like a big kid in a candy store when it came to helping the crew clean the Glock pistols, C-7, C-8 rifles, C-9 [light machine gun], assault shot guns and whatever else needed to be cleaned. I think he had to be held back from the .50 cal anti aircraft machine guns and the 40 mm cannons. I am sure ye old sea dogs now have your sea legs and bragging rights.

The atmosphere proved that this was a very exciting event for the town of Churchill and the Canadian Forces. The Rangers made a mark on both; the exposure will only pay dividends in the future for the programmes. Well done Churchill Rangers.

– 72 –

Rangers Receive Commander's Commendation

The Nickel Belt News, 5 December 2005

136 Canadian Rangers from Patrols in British Columbia, Alberta and Manitoba recently participated in a Brigade-level exercise, called Exercise Phoenix Ram, at the CFB Wainwright training area in Alberta. The exercise was the largest of its kind in Canada since 1992 and was designed to train Canadian forces for upcoming deployments to Afghanistan.

The training took place over a two-month period beginning in early September and ending in late October. 5,000 soldiers, mainly from 1 and 5 Canadian Mechanized Brigade Groups, as well as elements of the Canadian Air Force and support groups from across Canada, participated in training for what is called the Three-Block-War.

What has changed in today's areas of conflict, in terms of military tactics, is a greater departure from the older, more conventional types of operations to one that has a much broader focus. The overall direction in this type of operation is to deliver peace support and humanitarian aid, yet still be prepared to combat insurgents of collapsed regimes that threaten re-stabilization efforts. The Three Block War therefore requires soldiers to not only be trained in combat tactics in both a field and urban environment, but to be able to provide for civilians at the same time.

To train for such a deployment, one of the key ingredients of this type of operation requires individual soldiers and units as a whole to be very knowledgeable and very adaptable. The prime objective in Afghanistan is to help stabilize troubled areas and assist developing government institutions to rebuild the infrastructure needed to provide for the lives of the civilian population, and ultimately help the country to stand on its own.

The job of the Canadian Rangers involved in this exercise was to help provide the realism required to get as close to what deployment would be like in the theatre of operations. The exercise was elaborately scripted and required Canadian Rangers to play the part of civilians occupying three towns and thirteen farms. Troops entering these areas were confronted with a broad array of scenarios that ranged from ethnic quarrels and language barriers to civilian atrocities and seemingly everything in between.

As seen by day, Leopard tanks, mobile missile defence systems, armoured support vehicles, trucks and personnel carriers churned up the training area. Billowing clouds of brown dust followed in their wake and helicopters flew overhead. Insurgent units staged operations in and around the towns and farms and each of the 136 Canadian Rangers involved had a specific role to play, all of which was designed to challenge the capabilities of the many troops involved.

At times it was mayhem, when troops entered a village ravaged by fleeing enemy forces, while at other times everything was dead calm and every corner was suspect. In the distance F-18 aircraft dropped their ordnance on suspected enemy positions, the bright orange flashes could be seen before the muffled explosions reached your ears. It all felt very real and the importance of the task this training was designed for was only accentuated by the Canadian Ranger units portraying the civilians caught up in such a conflict.

That the Canadian Rangers managed to carry out their role with such authenticity is, in part, due to their involvement with CIMIC (Civilian, Military, Cooperation) units who have been deployed to Bosnia-Herzegovina and Afghanistan in the past. The commendation given to the 4th Canadian Ranger Patrol Group by Brigadier-General T.J. Grand said in part, "The Rangers provided assistance and tutored many of the other civilian role players who had limited experience in austere living. During the exercise, they spent many nights in the field in makeshift, often austere accommodation. They remained in character 24 hours a day adding the realism required to make the training and videos believable for the Primary Training Audience. The realistic training will definitely help all of the participants prepare for operations in Afghanistan. Throughout Brigade Training Event 2005, 4 Canadian Ranger Patrol Group demonstrated a combination of professional skills, innovation, dedication and cooperation that brought great credit to their unit and Land Force Western Area."

For their efforts, Rangers should be justly proud of their contribution to this very important and highly successful training exercise.

– 73 –

A Dream

Sgt J.-F. Gauthier

Bulletin Info Rangers (2 CRPG), December 2005

Cast off, speed 4 knots, direction Northwest. We are on our way to 8 days in the Great Canadian North. Ten rangers representing the various Quebec ethnicities (3 Inuit, 2 Cree, 3 Caucasians, 1 Montagnais, 1 Naskapis) and two instructors from 2 Canadian Ranger Patrol Group (2 CRPG) are about to discover the unique experience of living aboard the maritime coastal defence vessels Shawinigan and Glace Bay.

Departing from St. Anthony, Newfoundland, the vessels approach to pick up teams who will make up the crew. It's a first for 2 CRPG, as it has never before been in any way involved with the Canadian Navy. Objective: an inter-element operation that will facilitate the planning of future joint operations in order to maintain our sovereignty in the Great North. Indeed, the topic has recently been in the news, with the quarrel between Canada and Denmark over Hans Island.

After 4 days at sea, first stop Killiniq, the Inuit word for 'border' -- we are at the edge of Quebec and Labrador. We leave there in order to continue on to our next destination, Kuujjuaq, the capital of Nunavik.

The capital's population was awaiting us; the two vessels dock at the mouth of the river. After manoeuvring into place, the people are invited to visit the ships. Many of them take advantage of this opportunity to explore the vessels and question the crew about their work and their duties.

We repeat this experience 4 more times, in Quagtaq, Kangiqsujjuaq, Ivijivik, and lastly in Puvirniqtuq. At each place, the population was very friendly. Some senior citizens recalled that they used to see warships in their magnificent region. The last of these was the Protector in 1973, and never since have the Inuit seen such ships.

– 74 –

Wemindji Patrol Aids in Village Evacuation

Sgt S. Doyle

Bulletin Info Rangers (2 CRPG), December 2005

Summer had barely begun on the James Bay regions and already forest fires raged around the three Ranger communities of Waskaganish, Eastmain, and Wemindji. The first week of June 2005, saw an unprecedented amount of small to large forest fires occur. The village of Eastmain and the new hydro electric installation of EM-1 was evacuated to avoid endangering the members of the community and the army of workers on site.

The village of Wemindji, which sits north of the EM-l installation, had to deal with the secondary effects of such a large concentration of forest fires. The smoke haze that settled around the village aggravated some health problems in the community and the health board and the public security officer decided it would be better to evacuate elders, persons with respiratory problems and pregnant women to the safety of Val D'Or.

It was during the controlled evacuation that the Wemindji Ranger patrol was asked as citizens to help in the collection and control of movement at the airport to avoid any possible problems and to help those who required assistance. A safety cordon was established at all entrance points of the airport and once the plane landed a methodical procedure of embarkation was followed. No incident occurred during the evacuation process and all evacuees were safely loaded on the DASH-8 aircraft for their flight to Val D'Or.

– 75 –

Rangers From Across the North Assist Kashechewan

Sgt Peter Moon

3 CRPG *Newsletter*, January 2006

Canadian Rangers were a key part of the emergency assistance provided by the Canadian Forces to Kashechewan after the community's water was found to be dangerously contaminated.

Alarmed at the quality of the water, which had been under a boil-water order for two years, the Ontario Government ordered a mass evacuation for most of the settlement's residents, beginning at the end of October.

Aware that the province might order the evacuation, the Canadian Forces had already sent advance liaison teams into Northern Ontario to provide assistance in likely evacuation centres.

Kashechewan's Canadian Rangers were activated and took over the delivery of bottled water to residents. They also co-ordinated the transportation of most of the 1,120 evacuees who flew out of the community to emergency evacuation centres, picking them up at their homes and driving them and their bags to the evacuation aircraft. They also continued to provide bottled water for the 310 people who remained in Kashechewan.

Two Canadian Rangers from Kashechewan flew with the first evacuees to Sudbury, where they acted as interpreters and liaison staff for local authorities.

The work load in Kashechewan quickly began to overwhelm the resources of the overworked Kashechewan Rangers, who, in many cases, were working almost around-the-clock. Their first help came from Rangers from nearby Fort Albany who commuted daily to Kashechewan by freighter canoe until the Albany River began to freeze. After that, several Fort Albany Rangers flew to Kashechewan and remained in the community to provide help. Other Rangers flew to Kashechewan from other First Nations, including, Attawapiskat, Bearskin Lake, Constance Lake, Kitchenuhmaykoosib and Webequie. In total, more than 50 Rangers have helped in the Kashechewan crisis.

"In the first three days (before reinforcements arrived from Fort Albany) I'd have to say we all worked 20-hour days," said Sergeant Philip Stephen, the Kashechewan patrol leader. "When help came from other Rangers we really appreciated them coming. It gave us relief. The Rangers have done a really good job."

"The Rangers have been used for what they are trained to do," said Major Keith Lawrence, commanding officer of 3rd Canadian Ranger Patrol Group. "We've had Rangers involved from seven of the 15 communities in the North where we have patrols. That's tremendous. This is a good news story about

Aboriginal soldiers coming from across the North to the aid of an Aboriginal community in a time of need. It clearly demonstrates that we have the people and the capacity to do it."

The Canadian Forces' response to the emergency in Kashechewan included a military reverse-osmosis water-purification unit, known as a ROWPU, to provide safe drinking water. In ideal circumstances, the 10,000-tonne unit can produce up to 50,000 litres of safe drinking water a day. Several CC-130 Hercules aircraft, painted in drab camouflage-brown, flew from Canadian Forces Base Trenton to deliver the ROWPU and other equipment and supplies to Kashechewan. Engineers to assemble and operate the ROWPU were flown in from Canadian Forces Base Cold Lake, Alta., and Trenton. The big Hercules aircraft became a common sight at Kashechewan's small airport as they brought in regular shipments of supplies to support the ROWPU operation.

Canadian Rangers helped unload the planes and transport their cargo to the military water purification site.

Within 12 hours of putting everything in place, the ROWPU began providing safe drinking water from the tidal waters of the Albany River. The event was marked by a ceremony in which Archie Wesley, the First Nation's executive director, and Major Lawrence, the on-site military commander, shared two glasses of water to demonstrate its safety.

The water produced by the ROWPU was put in one-litre plastic bags and Canadian Rangers stored them in several buildings around the community to provide a strategic reserve of safe water.

"It's been incredible what's been done by the Rangers and by the rest of the Canadian Forces in the short time they've been here," said Mr. Wesley. "The people of Kashechewan appreciate what they have done for us."

In addition to the Canadian Rangers and the ROWPU, Major Lawrence said the Canadian Forces' response to the crisis in Kashechewan involved personnel from National Defence Headquarters in Ottawa, the Toronto headquarters of Land Force Central Area (the army in Ontario), military engineering units at Cold Lake and Trenton, staff from 3rd Canadian Ranger Patrol Group headquarters at CFB Borden, near Barrie, and 1st Canadian Air Division in Winnipeg.

"This is a tremendous story about the ability of the Canadian Forces to respond to the needs of a First Nation in difficulty," he said. "I'm very proud of the Canadian Rangers and of the composite military team we've put together. We've shown that the Canadian Forces can assemble disparate members of the Canadian Forces on short notice in a time of crisis, put together a good plan, and execute it well. It's a story about Canadians helping Canadians."

– 76 –

Fort Severn Rangers Help Make Military History

Sgt Peter Moon

3 CRPG *Newsletter*, January 2006

Canadian Rangers from four of Canada's five regional Canadian Ranger patrol groups have participated in a joint operation for the first time.

The historic event took place last August when HMCS Glace Bay and HMCS Shawinigan became the first Canadian Navy warships to venture into Hudson Bay in more than 30 years. Over a period of five weeks, the two coastal defence vessels visited several isolated communities in Quebec and Nunavut. They also visited St. John's and St. Anthony in Newfoundland and Churchill in Manitoba.

The ships carried and worked with Rangers over varying periods. The Rangers came from 1st Canadian Ranger Patrol Group, which covers the three territories, Quebec's 2CRPG, 3CRPG from Northern Ontario, and 4CRPG, which covers western Canada, including Churchill. Only the Rangers of 5CRPG from Newfoundland and Labrador, who worked with the Navy in a joint operation in 2004 along the Labrador coast, did not participate.

The ships were unable to sail as far south as Moosonee, so three Rangers and an instructor from Canadian Forces Base Borden flew from Fort Severn to join the ships during their community visit to Churchill. The Rangers were Sergeant Tommy Miles and Rangers Charlie Gray and Clifford Matthews. The instructor was Sergeant Wayne Atkins.

In Churchill, the four joined local Rangers, members of the ships' crews, air force personnel (who provided air support for the operation) and community members in a ceremony celebrating the Year of the Veteran.

Churchill is famous for its annual influx of polar bears which often wander into town while waiting for Hudson Bay to freeze. "The only bear we saw was hanging in a sling from a helicopter and flying out of Churchill," said Ranger Gray. The bear had been captured and was being removed from Churchill for the safety of the community.

The Fort Severn Rangers sailed from Churchill on board HMCS Shawinigan to the remote community of Arviat, Nunavut, where they met with local Inuit Rangers.

"When I joined the Rangers I never thought I'd go to sea with the Navy," said Sergeant Tommy Miles. "The ship was rocking pretty bad and I almost fell off my cot several times during the night. I won't forget it. It was the experience of a lifetime. They are awesome ships."

It was the first trip so far North of Fort Severn for all three Rangers but it was particularly meaningful for Sergeant Miles, whose grandmother is from York Factory. "She used to talk about it all the time and we flew over it," he said. "It almost made me cry to see it."

For Ranger Matthews it was also a memorable trip, "even if I couldn't sleep because the boat was rocking back and forth."

Sergeant Atkins said the three Rangers were impressed by the Rangers they met in Churchill and Arviat. "They couldn't get over the fact that both communities provided their Rangers with their own building and that both local patrols were highly organized and ready to be called out in an emergency," he said. "In Arviat they had a sled fully loaded, ready to go for search and rescue. I think they realized for the first time that the Rangers of Northern Ontario are part of a very large and professional organization. To be a Ranger in Churchill and Arviat means something. You are someone who is looked up to and respected."

The naval operation gave numerous Rangers opportunities to interact with other Rangers, said Major Keith Lawrence, commanding officer of the Rangers in Northern Ontario.

"The cultures of the 4,500 Rangers of the five Ranger patrol groups are very different, despite the fact they wear the same red sweatshirt and ball cap," he said. "From a Ranger perspective this has been a remarkable opportunity for the four Canadian Ranger patrol groups, really for the first time, to operate together.

"When it comes to the issue of sovereignty in the North it's important that you have federally-funded boots on the ground. Thanks to the Navy, it's been a true mission for the Rangers."

Combat Camera photo

The Canadian Rangers' motto is *vigilans* or watchful. A Ranger looks to the sea off Shingle Point, NWT, during Operation Beaufort in 2006.

Combat Camera photo

Ranger Ronald Minoza of Fort Simpson shows Corporal Devon Kidd how to set a snare during Operation Narwhal in 2007.

Combat Camera photo

Canadian Rangers and a CC-138 Twin Otter from 440 Squadron on Ellesmere Island during Operation Nunalivut 2008.

Combat Camera photo

Canadian Rangers Abraham Qammaniq and Simon Hiqniq load a snowmachine onto a kamotik with Sergeant Billy Budd on Operation Nunalivut 2010.

PW Lackenbauer photo

Canadian Ranger MCpl Jobie Oweetaluktuk and Cpl Allie Ohaituk on patrol at Witch Bay, along the Hudson Bay coast north of Inukjuak, Quebec, August 2006.

PW Lackenbauer photo

Canadian Rangers Daniel Oweetaluktuk and Jackie Inukpuk during annual training in Inukjuak, Quebec, in August 2006.

Combat Camera photo

3 CRPG played a pivotal role assisting with ROWPU operations, distributing drinking water, assisting community elders, and liaising with community members during Operation Canopy in Kashechewan during the fall of 2005.

DND photo

Canadian Rangers in Northern Ontario play a prominent role in an Aboriginal Veterans Day parade, 2008.

Combat Camera photo

Canadian Ranger Cheri Berlingette of the Zeballos patrol fires her rifle while Warrant Officer Dan Hryhoryshen and newspaper reporter Stephen Hume look on, 2009.

PW Lackenbauer photo

A Canadian Ranger navigates difficult terrain during Exercise Western Spirit, February 2009.

PW Lackenbauer photo

Canadian Rangers with 5 CRPG on the Labrador Sea en route to Hopedale, February 2005.

PW Lackenbauer photo

Canadian Rangers of the Cape Freels patrol on a tracker training exercise, October 2008.

1 CRPG Photo

Prime Minister Stephen Harper presents Charles, the Prince of Wales, with a set of red hooded sweatshirts and baseball caps to honour the prince's appointment as honorary head of the Canadian Rangers at a ceremony in Ottawa, November 2010.

DND Photo

Minister of National Defence Peter Mackay meets Canadian Rangers in Iqaluit during Operation Nanook 2008.

Combat Camera photo

Canadian Rangers Pauloosie Attagootak, Uluriak Amarualik, and Caleb Sangoya during Operation Nanook 2010.

– 77 –

Forces Plan Arctic 'Land is Ours' Mission: Armed snowmobile patrols hope GG will meet them on disputed ice

Adrian Humphreys

National Post, 9 February 2006

Canada's military is embarking on its largest affirmation of Arctic sovereignty, with five armed patrols snowmobiling 4,500 kilometres to converge in the High Arctic -- where the soldiers are inviting the Governor-General to meet them.

Scheduled to begin next month, the sovereignty mission over the Arctic islands and sea ice of the Northwest Passage is codenamed Operation Nunalivut, which means "land that is ours" in Inuktitut, the Inuit language.

The five patrols will pass through or near some of the Arctic areas that have been under increasing international dispute.

One patrol will be leaving from Ellesmere Island, off of which lies Hans Island, the barren rock claimed by both Canada and the Danes. That patrol will head west.

Three patrols will head east from Prince Patrick Island, which is to the east of an area of sea claimed by both Canada and the United States.

A fifth patrol will head northwest from Resolute over part of the Northwest Passage, an area that several countries claim is international water, but Canada claims is sovereign territory.

Several patrols will rendezvous on Lougheed Island, in the middle of the Arctic archipelago, on April 4, according to the plan provided to the National Post in a military briefing.

"The hope is that both the Commissioner of Nunavut and also the Governor-General will meet with the patrols on the ice at the rendezvous point," said Colonel Norm Couturier, commander of the Canadian Forces in the North.

A spokeswoman from Rideau Hall said Governor-General Michaelle Jean had not yet received the invitation, but would be pleased to consider it.

"This shows the flag and exerts our sovereignty. If you never show up there it is hard to claim your sovereignty," Col. Couturier said of the mission.

The patrols are scheduled to begin their trek on March 28 and return on April 9.

Each patrol will consist of about 10 members, two regular force soldiers and eight Canadian Rangers, a largely aboriginal reserve unit based in the north.

"We're going to be covering a larger piece of the north than has ever been covered before in a patrol of this kind," said Captain Conrad Schubert, a spokesman for the 1st Canadian Ranger Patrol Group (1CRPG).

"There have been a lot of logistics to sort out and it is all nicely co-ordinated on paper. Now we have to see how it co-ordinates on ice," he said.

These missions are typically difficult and sometimes dangerous. The patrols will be travelling on snowmobiles over brutal terrain, often over sea ice, in temperatures of -40C. Two of the patrols will traverse Polar Bear Pass on Bathurst Island, a channel between two mountain ranges. The patrols will be armed with shotguns for predator control.

The rocky terrain in the far north has wrecked several snowmobiles on previous patrols, and accidents have also forced emergency airlifts of injured soldiers.

For the first time, the patrols will be linking up with one another -- a difficult task when little of the area has been accurately mapped and visibility can be poor.

"It will be challenging, but it is feasible," said Major Chris Bergeron, commander of the 1CRPG.

"The patrols will meet each other in the middle of nowhere on the ice so we can prove to the entire world that we can deploy Rangers anywhere and they can meet and can react to any major air disaster or any operation they request of us," said Maj. Bergeron.

Along with the sovereignty mission and the exercise of linking small, autonomous patrols into larger groups, the operation will also document the existing infrastructure of the north -- old wartime airfields, abandoned weather stations and civilian and military exploration outposts.

"There is a fair amount of semi-abandoned infrastructure in the Arctic that is still serviceable but not currently used," said Capt. Schubert.

At Mould Bay, for instance, there is an abandoned Environment Canada weather station with living quarters and a transport shed where snowploughs and trucks remain.

The soldiers will try to restart the equipment and the station's generators after years of remaining frozen, silent and still.

"This is an opportunity for us to see what the problems are operating out of a structure or station that has been abandoned for many years" said Capt. Schubert.

"From what I understand, coffee cups with dried coffee stains and the remains of cigarette butts are still on the table where the last crew left them when they departed. I'm told the calendar on the wall is from 1995."

An accurate inventory of infrastructure could aid in responding to a crisis in the north, such as an airliner crash or military threat, said Maj. Bergeron.

The mission was planned in November and approved in December, military officials said.

Current satellite images are being used to map the patrol routes and fuel caches and food rations are already being deposited along the routes by air.

It is likely to be the most expensive sovereignty mission so far, as well as the largest.

Previous, smaller patrols have cost hundreds of thousands of dollars each. A 2002 patrol of 34 soldiers to the magnetic North Pole cost $594,803; a 2003 patrol of 11 troops to Banks Island cost $336,097; a four-man patrol to Alert, the world's northernmost permanently inhabited settlement, in 2003 cost $204,894; and a 20-man visit to Alert in 2004 cost about $700,000, according to documents obtained by the Post.

Maj. Bergeron expects this mission to cost $1-million.

This the first patrol under the name Operation Nunalivut. The new name carries stronger symbolism. Previous patrols were codenamed Operation Kigliqaqvik -- from the Inuktitut word for "the place at the edge."

The five patrols will also be testing a new satellite tracking device to allow military headquarters to more accurately track their progress and also test ground-to-air communications in the deep cold.

– 78 –

The Guys Who Went Out in the Cold

John Thompson

Nunatsiaq News, 10 February 2006

When a bunch of soldiers from the South try to survive in the Arctic, they screw up their igloo and catch fish too small to eat.

The lumpy-looking snow house stands next to another, far more orderly igloo. To be fair, the first started as an oval, while their rivals used a circular design. But now the half-finished oval has acquired a tilt, its walls rising and falling chaotically.

"Elders can do it in about half an hour. We've been doing it all day. It gives you an idea of their skill," says Master Cpl. Eric Viau, who smoothes another snow block on the wonky walls. Around him, fellow soldiers are placing bets on whether their igloo will stand by day's end.

Viau is one of 84 soldiers from Gagetown, N.B., gathered outside Iqaluit at Iqalugaajuruluit, past Tarr Inlet, last Friday for a sovereignty operation dubbed "Glacier Gunner."

It's a fitting title for a mission that saw their powerful 50-calibre machine guns freeze, then break during firing practice.

Good thing the soldiers are accompanied by 11 Canadian Rangers. It's no coincidence that the wobby igloo lacks the guidance of an Inuk, while their symmetrical rival has Sgt. Dinos Tikivik inside it, reaching to grab ice blocks passed to him by a string of soldiers.

"They've come a long way," Tikivik says of the troops. "They only knew how to set up a tent."

During three days the soldiers will build igloos, ice fish and hunt seal and caribou. Or at least they'll try.

The morning began with Sgt. Brad Young clipping a parked snowmobile as he pulled up to Iqaluit's Gas Bar to meet the mayor and escort her to camp. He's apologetic, but doesn't need to be — he won't find much argument with a C7 assault rifle slung across his shoulder.

He lists off some lessons already learned.

Lesson one? Don't stand still when it's -25 C. Armed Forces personnel are used to loitering about as they wait for orders. It's a good way to freeze. When Rangers notice the troops shivering, they tell them to run around their snowmobile and qamutik. Or they just wrestle them to the ground.

Lesson two: Look where you leap. When the soldiers' snowmobiles bogged down in overflow slush, Young says their first instinct was to jump off and push.

Bad move — their legs were soon soaked, and by the time they returned to base, had begun to freeze.

When a female Ranger saw this, Young describes how she took one man's leg, pressed it against her bare belly and leaned forward. "That captures the spirit," he says of the relationship between the soldiers and Rangers.

The Rangers' Tikivik says the military's also learned from Operation Narwhal during the summer of 2004, when two soldiers became lost in fog and spent a night by themselves before they were rescued.

"For here, if we're going anywhere, it's with two Rangers," Tikivik says.

In return for the mutual aid, Rangers had a chance to fire off the guns used by the soldiers. Well, they did before they froze and broke.

A 15-minute snowmobile ride away, another platoon learns another lesson: Hang on to those military rations. The soldiers cluster around holes bored into the ice, hoping to hook a fish. Gunner Eric Landry was among the lucky: he caught one after about six minutes of jigging. Between the dozen-odd men, only five or six tiny fish have been caught.

"Not enough to feed an army," says Gunner George Spilkin, who wasn't so lucky and gave up after 10 minutes.

Back at base, the Rangers call the little fish "sardines."

The soldiers have had to unlearn some southern training, like never put snow on frostbite. The Rangers say otherwise, arguing this lessens the shock to frozen tissue as it thaws. Ice-numbed soldiers seem to agree.

Much of the soldiers' education involves how to stay warm. Before camping out, soldiers spent a night inside the FOL as they tried to adapt to the cold. The following evening, out on the land, a few others froze — until they followed Rangers' advice and dug a cold trap at their tent's entrance.

But they haven't had to worry too much about the cold, with relatively mild weather of about -25 C most days, with winds pushing the temperature down to -44 C one afternoon out on the ice. Still, four soldiers on Friday are being treated for frostbite back in the FOL hanger while the others try their hand at survival skills.

Lesser lessons abound. It turns out the military-issue gloves, Thinsulate with leather grips, don't keep out the cold as well as simple cotton workgloves that cost $1.50 from the Arctic Survival Store.

Snowmobile and sled tips include: don't use the gas you mixed down south. It will foul the snowmobile, causing it to splutter and stall.

Oh, and you can forget about Stephen Harper. This operation was planned far before the recent federal election campaign, which saw Canada's new prime minister promise to beef up our country's military presence in the North. Far from television sets, the soldiers only become confused when asked about how their igloo-building skills fit into the next Prime Minister's military scheme.

As the soldiers mill about on the ice, they look a little silly in their olive green uniforms, intended to provide camouflage in a forest. They do pack "whites," but they won't be used in this operation. Rangers are known to use the white uniforms to sneak up on seals.

By dusk, one igloo stands complete — the one the Rangers worked on. It's big enough to hold 25 people inside, and some troops are talking about sleeping inside it. As for the wobbly one — it's been given up as a lost cause.

During past operations, Rangers and other troops would sleep in different tents. Not this time — as each soldier prepares to bunk up, he can count on sharing a tent with a Ranger. That means if any of the Gagetown soldiers have questions about keeping warm, which they will, they'll be sure to get an answer. Or at least a wrestling match.

– 79 –

Ranger Trek Comes to an End

Andrew Raven

Northern News Services, 17 April 2006

When Alan Pogotak pulled into the ice-covered harbour of Resolute last weekend, his face was covered with polar bear kisses.

The dark patches on his weather-beaten cheeks came from two weeks of travelling across some of the harshest terrain on the planet - part of an ambitious mission to stake Canada's claim to the High Arctic.

"I'm glad to be back," said the soft-spoken Ulukhaktok (Holman) native. "The first thing I'm going to do is take a shower."

About 50 Rangers and armed forces personnel received a hero's welcome when their column of snowmobiles pulled into the small bay off Resolute Sunday afternoon.

The Rangers were among the best from the Northwest Territories, Nunavut and the Yukon.

Their mission, code-named Operation Nunalivut, was designed to test Canada's ability to operate in the High Arctic. The Rangers crossed more than 5,000 kilometres of sometimes brutal terrain, weaving their snowmobiles across sections of jagged sea ice that broke skis and splintered wooden sleds.

The patrol concentrated on the western islands of the High Arctic, a huge swath of uninhabited land that stretches from Resolute on Cornwallis Island to Prince Patrick Island. The area is so far north, it was home to the ever-shifting North Magnetic Pole in 1999.

"This (was) by far our most challenging exercise," said Lt-Col Drew Artus, Chief of Staff for the northern wing of the Canadian Forces. The region is dotted with abandoned buildings, airstrips, oil wells and test-mines.

In an area with almost no infrastructure, the meagre remains are a bonanza for the Canadian military. Part of the Rangers' mission was to catalogue what was out there.

Some of what they did find, they used. Several Rangers and soldiers weathered temperatures that dipped to -40C in an abandoned Environment Canada weather station, while pilots landed Twin Otters on airstrips built decades ago by prospectors.

The mission comes while critics question whether Canada has the money and manpower to keep tabs on the Arctic, a region that is home to a wealth of resources including gold, diamonds, oil and natural gas.

Some believe Canada's claim to those riches could be jeopardized without a strong presence in the North -- something Nunalivut was designed to address.

"It is important for us to go up there and show we can do things," said Artus. "It says this land is ours."

While there is no immediate threat to Canada's Arctic sovereignty, missions like Nunalivut send and important message to the international community, say military officials.

"If we don't go to Isachsen - a remote weather station - for two months, will the Americans plant a flag there? No. But we have to be able to operate in the North."

For about $100 per day, the Rangers braved treacherous sea ice, brutally cold temperatures, raging blizzards - including one storm where visibility was reduced to a couple of metres - and polar bears.

"It was a fun trip," said Repulse Bay's Merrill Siusangnark, after arriving in Resolute. "But it feels good to be back."

That sentiment was echoed by Hall Beach's Ike Angotautok. "I'm a little tired of the cold," he said.

The Rangers, who travelled by snowmobile, covered more than 100km on clear days. Sometimes the Arctic ice was as flat "as the Collisee in Quebec," said Major Chris Bergeron, commander of northern Ranger units. But rough sections and storms could ground the patrols to a complete halt.

The military took over a Co-op store house - which was originally built for exercises like Nunalivut - where they set up communications gear to keep in touch with the Rangers as their five separate patrols weaved across the Western Arctic.

With years of experience living on the land, Bergeron said the Rangers are important part of Canada's defence plan for the North. "I learned so much in the last 15 days from them," Bergeron said Sunday after reaching Resolute. "Like how to light a stove at -50C in a blizzard."

Despite temperatures so cold "you could not have any exposed skin", the Rangers emerged from trip more or less unscathed, said medic Rick Mackwood.

The cold did play havoc with medical supplies and Mackwood had to wear rubber surgeons' gloves over his mittens in the field.

The military budgeted $1 million for the exercise, but poor weather drove up the cost at least an extra $500,000, Artus said.

There are 1,700 Rangers in the three territories, according to the military.

Created in the aftermath of the Second World War, their job is to keep an eye on a 300-km circle of land around their home communities and - about once a year - embark on patrols like Nunalivut.

With only a handful of permanent bases in the North and shortage of satellites concentrating on the polar region, they have become the eyes and ears of the army in the North.

– 80 –

Manitoba Canadian Rangers Support Exercise Charging Bison, Winnipeg, May 2006

The Ranger, Spring 2006

"We want food! We want blankets! We want water!" Between the angry shouts came the calm, clear voice of the refugees' leader, Ranger Chris Hart, from the Churchill Patrol, "I don't know how long I can hold them back."

In this scenario, twenty Canadian Rangers set out to simulate a rampaging mob outside a wired encampment of Canadian soldiers in downtown Winnipeg. They had been displaced by an unfriendly warlord and told to shove off without food, water or blankets into the chilly prairie landscape. Chanting their needs to the suspicious soldiers safe behind their wire in a secure Alamo, the Canadian Rangers finally received some water. Food and blankets were not forthcoming so the more hotheaded among them tried crashing the gate. A very tall [Displaced Person] among the refugees, clad in his yellow slicker and looking suspiciously like Captain [Doug] Colton, succeeded in removing the pins from the top of the fence and the gate, creating a gap. Before the mob could move into the breech, a back-up force of Canadian troops ran over and kept them from charging through. Driven back, they resumed their protest under the watchful eyes of two Winnipeg policemen who were stationed onsite to ensure the public did not get the idea the scenario was real. As the mob withdrew, one of the policemen made history with a comment likely never before uttered by a cop to rioters when he said, "Nice riot. Good job, everyone."

Bedraggled, cold and hungry, although most of them looked surprisingly well fed to be masquerading as displaced persons, Canadian Rangers participated in Exercise Charging Bison 2006.

Sometimes they were homeless refugees, sometimes-angry citizens of an enemy-occupied city, sometimes terrorists: always Canadian Rangers.

And, once again our Canadian Rangers made a successful contribution to a Canadian Forces training exercise. Twenty Canadian Rangers from Patrols in Manitoba recently participated in a Brigade-level exercise, called Exercise Charging Bison 2006, in the city of Winnipeg. The Canadian Rangers arrived on 29 April and stayed until the 7th of May. They were quartered at McGregor Armouries in the north end of Winnipeg. The exercise, the first collective training event for 38 Brigade's task of generating a force of 200 plus soldiers in support of overseas duty in 2008, required the Canadian Rangers to participate as terrorists, refugees and citizens of a nation under siege. Winnipeg was re-named, "Windahar," a city in the failed state of West Isle. The media report tells us that a dishonest government and a civil war between warlords has put the city into a state of chaos and disorder. There are three major groups busy carving up their own chunks of the empire known as "Windahar." They are the Internal Solidarity Movement (ISM), an active rebel group known for sniper attacks, ambushes and improvised explosive devices (IEDs); the People's Liberation Army (PLA), which is a group of local militias, friendly in some areas but hostile in others and the Black Dawn Brigade (BDB), a militant, extreme collection of bandits believed to be in support of al-Qaida. The Black Dawn Brigade uses suicide bombers and wants Canadian soldiers dead. Depending upon the time and place the Canadian Rangers acted as members of all three of those groups. Captain Doug Colton, the Manitoba Det Commander, arranged for Canadian Rangers from Gillam, Lynn Lake, and Ste. Theresa Point to fly in and support the exercise.

Five hundred soldiers from 38 Brigade, which included Manitoba, Saskatchewan and the Thunder Bay area of Northern Ontario, participated in training in this citywide exercise. The aim of Charging Bison was to train soldiers in the Three Block War concept where Canadian troops on deployment could become involved on one city block in humanitarian assistance tasks, one block further along they might be required to render support in Peace-keeping roles and then one block further might even encounter the need to participate in a combat operation. In a theatre such as Afghanistan, soldiers need to be proficient in all three areas and have the skills and training to distinguish one from the other. Decisions, right down at the individual private soldier level need to be made instantly.

All week long as the soldiers learned lessons in an urban environment in Three Block War, the Canadian Rangers were employed as displaced persons, irate citizens, wounded or injured terrorists from a country that had Canadian troops trying to lend assistance. In many cases, that assistance was not welcomed by the (Ranger) citizenry. Section level security patrols, marching down city streets were frequently met by hostiles who wanted no part of foreign soldiers from faraway Canada. The Canadian Rangers found that putting this pressure and harassment on the soldiers demonstrated how well our Reservists and Regulars were able to take this verbal abuse.

In a couple of scenarios, in order to add realism, the Canadian Rangers as "citizens" even requested aid in a foreign tongue. Soldiers on patrol were stunned when they heard requests for information about a lost family member, pleas for food, water and shelter being yelled at them by displaced persons. Normal enough for the exercise, but what had bewildered the soldiers was that all of these requests were coming at them in the Cree or Dene language.

Early one morning, the Rangers, acting as terrorists, holed up in a deserted warehouse at Kapyong Barracks in the southern end of the city. They had taken hostages and were all set to withstand a siege. Out of the warm sunny skies of Friendly Manitoba streamed eight helicopters, heading for a field 500 metres to the north of the warehouse. Landing, they spilled out a platoon of Canadian and US assault troops, who formed up and charged the building.

Simultaneously, another force of soldiers dismounted from trucks and established a cordon around the building. Once they were in place, the troops from the choppers charged the warehouse, blasting their way in through the front doors. As the troops moved to clear the building room by room, Exercise Controllers followed up, taking Canadian Rangers and placing them in positions where the POW teams could find them and deal with them. Other Canadian Rangers were placed where the medical teams and the clean up crews could remove them to the outside area. A short firefight and the battle was over. Once the warehouse was secured, the hostages freed and the prisoners led or carried away, the medical teams took over.

The Canadian Rangers, very convincingly made up to demonstrate a variety of battlefield wounds, were carried or escorted from the battle scene. With lessons learned from Vietnam and the Gulf War, the medics evacuated the wounded to a MASH-like tented hospital. There, triage teams steadily and efficiently worked to save the wounded soldiers and terrorists. So effective and realistic was the make-up for the wounds that a real casualty stood by the hospital tent door waiting, his hand wrapped in a bloody bandage, to let the stretcher bearers in first. When he was told it was part of the exercise, he said, "It looked so real that I felt my cut fingers could wait."

The next encounter was between three Canadian Rangers from Ste. Theresa Point (Rgrs Agnes Mason, Curtis Mason and Jemima Manoakeesick) masquerading as a mother, a son and a daughter searching for another daughter lost during some artillery shelling. Wandering through a field they came across an Artillery unit training with their guns. Quickly they were confronted by the perimeter sentries.

Things got very interesting and frustrating for the Canadian troops when the little family requested help. Interesting because while the Canadians wanted to help, they couldn't understand the language being spoken to them. It got even more suspenseful when two hangers-on approached. One of them, Captain Colton, no doubt encouraged by his fence-destruction at the previous site investigated things even further. He insisted in broken English that the soldiers participate by getting a search party up to look for the missing person. I was the second hanger-on and when I pulled out my trusty camera, all hell broke loose. The Gunners all began yelling to grab the camera, confiscate the camera or just plain shoot the photographer. The friendly forces had suddenly stopped being friendly. Even the controller armband I was wearing to indicate that I was not part of the exercise, did not slow them down. I shoved the camera in my pocket and then played mute.

The soldiers, this time with no fence between them and their Canadian Ranger opponents as had occurred at Pine Arena, became a little pushy. Giving Aggie Mason a shove the soldiers got even more excitable and when one of them gave Captain Colton a little push he shut the scenario down. We explained the situation to the soldiers and all was well. With darkness coming on we made quick trip to Tim Horton's for coffee and hot chocolate.

Every day at least one group of Canadian Rangers was trucked out to the St. Boniface area of town. There, usually, under a dull and leaden sky, accompanied by the spring drizzle, they wandered the streets in search of the despised Canadian foot patrols. Once they ran across a patrol, these angry citizens of a city held hostage by invading troops proceeded to make the soldiers' lives miserable. "You soldiers, you leave our city." And "Go back where you came from!" were interspersed with "Soldier! You give us cigarettes?"

Often the insults and catcalls were issued in Dene or Cree and that created even more confusion. The soldiers on patrol were trying their best to forge links of friendship, help where they could and if necessary, assist where they could. To have to face a small crowd of hostile citizens did not make their job any easier.

Sometimes the foot patrols were accompanied by G-wagons. Built by Mercedes-Benz, these LUVWs or Light Utility Vehicle Wheeled carried four soldiers and their kit. They have already seen service in Afghanistan. Here in Winnipeg they followed the foot patrols and offered more firepower, should the need arise.

During the early days of the exercise, the Canadian Rangers had the added help of some real protesters, upset at Canada's decision to even be in Afghanistan. Over a hundred protesters attended one scenario and once on site, some of them handcuffed themselves to bridge abutments, others followed the troops on bicycles and yelled at everyone about how disgraceful it was to be doing your duty to your country. The police arrested a number of them, the soldiers patiently ignored them and then the protesters, one by one, gradually faded away, leaving the field of protest to the Rangers.

Day three, or more like night three and darkness comes in fast when you're in an unknown part of the city and your walkie-talkies have quit working. I was driving three Canadian Rangers to an RV where we could meet up with WO Chris Fink, the Chief Clerk of the Manitoba Det and her vanload of aggressive citizens and then participate in a vehicle checkpoint. Since the car I was driving was a rental, we thought the soldiers at the checkpoint would let me drive through without any questions being asked.

We blundered around, all of us offering directions or cursing the walkie-talkies. Giving up on ever finding WO Fink and her vanload of Ranger terrorists we decided to turn back to what we had thought was the meeting place.

"Stop the car!" "Put out your headlights!" "Get out of the car!" and other assorted loud bellows came from all around us. We had found the checkpoint. Since it was a rental car, I had no idea how to turn off the lights. Until now they had shut themselves off whenever I turned off the car. Our best bet, I thought was to back up and get out of there. Lights burst into my rearview mirror and added to our problems. WO Fink and her cavalry had arrived. They pulled in behind us and waited for the outcome.

Thinking I was driving my own truck, I tried to pull over to the side of the road, but I was still in reverse. The soldiers now alarmed that we were making a getaway, yelled louder and issued more irrelevant demands. I shut the car off and the lights did go out. But now I was unable to move it from where it had come to rest; diagonally in the road.

Yielding to their demands, I got out of the car to explain that I was part of the control staff and not subject to the rules of the exercise. No one wanted to hear that, so I was ordered to put my hands on the hood and stand still. Finally an Exercise Controller appeared and we agreed that I was there to take pictures, not be strip searched in a dead-end, darkened, Winnipeg street.

The scenario played out with the Canadian Rangers who had been in my car having to undergo that search. WO Fink and her gang of ruffians sat in their van, yelling at the foreign devils who had taken over their city. Once again those Canadian Rangers who were bilingual, used Cree to insult the soldiers. The Exercise Controller thought this brought an excellent touch of realism to the situation and let them keep yelling. But because we had six or seven people in the van it would have taken until midnight to complete a search of all of them, he called off the strip search when the three Canadian Rangers from my car were finished with.

Normally the activity is played out and then, once completed, the soldiers were debriefed by the Exercise Controllers. Our vehicle checkpoint phase needed to have a kind of halt brought in due to one of the soldiers losing his cool and yelling and swearing at us as we sat in our vehicle. The Exercise Controller took him aside and explained that he needed to settle down, act in a more soldierly manner and be polite as up until they found any incriminating evidence, we were just ordinary citizens out for an evening ride.

The exercise ended too soon for the Rangers, who told me that they had enjoyed playing the roles they had been given. One Ranger said that it was a valuable opportunity to mingle with the Reservists and Regular Force personnel participating in the exercise. Another explained that while they were only acting in a variety of enemy actions, that they were fully aware of the importance of "playing the game."

All of the scenarios were designed to recreate the real thing. As Captain Scott Carlson said in the Winnipeg Free Press, "You can't do this unless you use real scenarios. It was to be as real as you can get so the training is maximized." He added, "At the end of the day, some of these people will be deployed."

The Canadian Rangers understood this and as one said to me, "We feel very useful doing what we are doing here. There's not a lot we are learning that will add to our Ranger skills, but that's not as important as knowing that these soldiers will be going off to Afghanistan in 2008. We may have helped them stay alive."

– 81 –

Rangers Take to the Water Inspecting Canada's North Warning Radar Sites

Sgt. G.J. (Junior) Roberts

The Labradorian, 5 September 2006

The Canadian Rangers embarked on an important and unique journey last month as the group carried out an inspection of four North Warning Radar sites along the coast of Labrador.

On Aug 03, 2006 the Canadian Naval Vessels HMCS Montreal, HMCS Moncton and HMCS Goose Bay sailed from St John's for Canada's far North. Onboard the latter two were Sgt. Tony DuBourdieu, a Canadian Ranger Instructor from the Gander Detachment Headquarters, Ranger Levi Nochasak, Ranger Sgt. Henry Broomfield, Lt. Joe Anderson, Ranger Sgt. Dean Coombs and Ranger Albert Tuglavana.

The Rangers on the Goose Bay and Moncton fell easily into the ships' routine as they proceeded towards the Northern Labrador Coast, delayed only by a quick stop in Nain for Lieutenant Governor Ed Roberts' visit (he was travelling on the HMCS Montreal). All enjoyed the celebration in honour of the Lieutenant Governor's visit, which included traditional entertainment (drumming, singing and dancing) and other festivities.

The journey continued, sailing north to inspect radar installations that are part of Canada's North Warning System.

After being dropped at designated areas along the rugged North Labrador coastline, the Rangers and ship's crew faced a long walk (approximately nine km. round trip in most cases) up a very steep, rugged mountainside, sometimes encountering snow banks that created a fog-like vapour and in all cases having to traverse a terrain that, combined with altitude, was a decent workout even for those in good physical shape. However, the view from the top and numerous sightings of wildlife proved to be worth the walk.

All sites are equipped with cameras, and communications with North Bay had to be established as soon as the group arrived at the sites. The inspections revealed very little in the way of damage, a door open here or a lock broken there, but nothing that would cause any alarm. The radar installations on three of the sites have been visited, but weather conditions demanded a halt to the advance on one site, allowing only for inspection of the base facilities.

With the mission completed, the HMCS Goose Bay and HMCS Moncton continue North to Iqaluit (numerous icebergs and a Polar bear on Baffin Island enroute all make for an interesting passage).

At Iqaluit, the Rangers bid farewell to the ships' crews. The following morning, the Rangers flew home. All three ships continue on their northern journey to complete their mission, Operation Norploy.

The 5 Canadian Ranger Patrol Group thanks the crews of the HMCS Moncton and HMCS Goose Bay for their assistance in enabling the inspection of Canada's North Warning Sites. Special thanks as well to Sgt. DuBourdieu and the five Canadian Rangers for carrying out the inspections.

– 82 –

The Critical Role of the Canadian Rangers

Kerry McCluskey

Naniiliqpita, Winter 2006

As Abraham Kublu grew from a boy to a man, he watched his grandfather Solomon Koonoo hunt on the land and sea near Pond Inlet. Today, the expertise he developed, which has been passed from one generation of Inuit to the next for thousands of years, is in high demand with the Canadian Forces.

Aboard the Canadian navy warship HMCS Montreal this past August, Kublu was headed to an observation point on Bylot Island in Sirmilik National Park as part of Operation Lancaster. He was one of 16 Canadian Rangers from the Pond Inlet Ranger Patrol participating in the military exercise executed by the Canadian Forces' Joint Task Force (North) to exert sovereignty in the Northwest Passage and other Canadian internal Arctic waters by ship, aircraft and land troops. It was Kublu's first time aboard a navy ship, and his first time taking part in a sovereignty exercise.

"I get a little tickle in the back of my neck when I think about them depending on us," said Kublu. "This is important to Inuit because we've never had such military presence here. It makes us feel more like Canadians."

As is the case when the military conducts events in isolated, coastal regions that lack the presence of other Canadian Forces, the Rangers provided critically-needed local expertise. Called Siuraliq, meaning there's sand, the observation post where Kublu was based is an area he knows well from years of winter polar bear hunting and spring narwhal hunting. This is exactly the sort of familiarity the military is after.

"We're watching for polar bears, we're watching the water, and on our free time, we want to do some hunting for narwhal and caribou. We know the area and the vegetation of the area and where there's a good chance of catching animals. We have more expertise, and we know if it's dangerous or safe," said Kublu.

This element of added safety is critical to lending a sense of ease to such Arctic missions.

"It can be intimidating when the nearest community is 400 nautical miles away, but the confidence you get from having the Rangers with you as part of the patrol puts you at ease," said Captain Brian Wiltshire, the Deputy Commanding Officer of the First Ranger Patrol Group. Wiltshire also sailed on the HMCS Montreal during Operation Lancaster and oversaw the participation of the Rangers. "The Rangers are very valuable because of the knowledge they bring to the patrol and the operation. They have the knowledge from the Elders on where they should and shouldn't go, if it's hunting areas of spiritual areas… so you don't interfere with the herds or things like that," he said.

Being able to depend so heavily upon the Rangers means that the Canadian Forces don't consider heading to the Arctic without the helping hands of the local patrol.

"Whenever we take Canadian Forces to the North, they must be twinned with a Ranger patrol. Survival in the Arctic is not necessarily something the forces from the South are used to. Rangers guide them in traditional ways. They help them out in their own mission and they extend the culture of the North to help them connect with Canadians," said Colonel Chris Whitecross, the Commanding Officer of Joint Task Force (North) and the lead on Operation Lancaster. "We don't send any troops up North without the Rangers' help," she said.

This strong dependability upon the Rangers is not lost on Nunavummiut. Nunavut Commissioner Ann Hanson, also aboard the HMCS Montreal, described the Rangers as vitally important and said their knowledge of the land, sea and skies inspired deep respect from Inuit in Nunavut.

"Every time I go into a community, I see the respect and admiration of their peers. They have the skills for survival," said Hanson.

Of the 58 Ranger patrols that exist across Canada's North, 25 of those are in Nunavut. The territory's 761 Rangers regularly conduct surveillance during sovereignty patrols, provide guiding expertise, and teach the military survival skills like igloo building, hunting, fishing, navigation, and dog team travel. Rangers must be at least 18 years of age, in good physical health and have good knowledge of the land.

Nunavut may soon benefit from an increase in the number of Rangers. Whitecross said that federal Defence Minister Gordon O'Connor stated he intends to increase the Ranger program across Canada by 500 people, but Canada Command, the entity that controls the Canadian Forces, has yet to determine how this increase will be dispersed across the county.

At a press conference in Pond Inlet on the heels of Operation Lancaster, Nunavut Premier Paul Okalik said he appreciates the promised growth of the Ranger patrol, but would like to see Inuit involvement in sovereignty matters progress further. Okalik said he supports the federal government's push to assert their sovereignty in the Arctic because if Canadian internal Arctic waters were deemed international waters, "Inuit would have no control over what would happen in those waters. That's a major concern."

About to feast on fresh Arctic hare at his observation post in Dundas Harbour, some 248 kilometers north of Pond Inlet, Ranger Norm Simonie said his priority during Operation Lancaster was to do his part to safeguard the land, sea and animals, and to ensure all the cruise ship passengers dropping by knew the area was part of Canada.

"We hunt here so I want to keep this as ours. This is our hunting area for muskox, walrus, beluga, polar bear, rabbits," said Simonie.

Brian Simonee agreed. "It seems too easy for anyone to come in and do what they want in our water. We have to show them that this ours," said Simonee.

– 83 –

Home on the Range: Along our wild and remote coastline, a military unit with roots in the Second World War soldiers on

Stephen Hume

Vancouver Sun, 3 February 2007

Nootka Island—Far to starboard, precisely where the oyster grey sky melted into a slate grey ocean, dazzling columns of white suddenly punctuated an otherwise featureless horizon.

Those jets foam erupting skyward signalled exactly where heavy swells from the open Pacific seethed across the barely submerged teeth of Nutchatlitz Reef, a maritime hazard that earns a major caution in the government's *Sailing Directions.*

On the afterdeck of the small boat *Perseverance,* which skipper Greg Brooks was deftly turning across the seas about 300 kilometers northwest of Vancouver, stinging sheets of spray came flying over the rail.

As the spindrift rattled loudly against the stiff tarps that kept the 25-kilogram packs and other gear dry, Canadian Ranger Cheri Berlingette just shrugged deeper into her flotation suit and grinned.

The object of her subdued amusement was Canadian Force Warrant Officer Dan Hryhoryshen. The career soldier was looking a tad green about the gills as the boat rolled. In fairness, Hryhoryshen, the regular forces instructor on this military exercise off the remote outer coast of Vancouver Island, had confessed to me his lifelong propensity for seasickness before he boarded. For that matter, I reminded him that Admiral Horatio Nelson made frequent trips to the rail, too, but he still won the battle of Trafalgar.

Pte. Berlingette, of Zeballos, on the other hand, works on a halibut boat and was plenty used to dirty weather and the queasiness induced by an unpleasant combination of engine fumes and the corkscrew motion of a pitching deck.

Brooks, 53, who was also grinning at Hryhoryshen's discomfort, is a former patrol commander for the Rangers and after 40 years in Zeballos is prized by the regular forces for his knowledge of the woods and water.

This was their natural environment, so I suppose they could be forgiven for taking the silent Mickey – in Brooks' case, not so silent – out of the professional soldier whose own superior, Capt. Jim Murray, had described for me in expansive superlatives before we left.

"I've never met a guy with his knowledge of the ground," Murray said of Hryhoryshen, whose job it is to provide hands-on leadership and instruction in the arcane skills of reading topographical maps, handling the Rangers' venerable Lee-Enfield .303 rifles and communications using standardized military procedures to avoid confusion.

The Fourth Canadian Ranger Patrol Group, headquartered in Victoria, has 23 patrols in 26 small communities across B.C. and musters 390 Rangers, 70 of them women like Berlingette and Leanne Humphrey, a youth and child care worker at the school, acting mayor and Zeballos councillor.

The Rangers are an extension of the militia and descend from the founding of the Pacific Coast Militia Rangers in the spring of 1942, the darkest days of the Second World War when West Coast landings by Japanese forces where thought a real possibility.

It dawned on the highest command that B.C.'s 8,000 kilometers of coastline, comprised of countless islands, treacherous currents, hidden reefs, shore bogs, trackless tangles of cedar and vicious thickets of devil's club would be impossible to secure using conventional methods.

"The country is for the most part rough and rugged, it is very sparsely populated and much of it only known to those whose employment or recreation takes them away from settled areas and off the highways.

"It is only by means of an organization such as the P.C.M.R. that possible enemy activities can be countered effectively," an official military brief reported to Ottawa.

Local Knowledge

Who better to patrol it than the people who lived there, could read the weather in the wind and the sky, knew the terrain intimately and could find their way through the dense fogs, pelting rains and dense underbrush.

Within a few months, 4,000 loggers, timber cruisers, fishermen and native Indians had been recruited as Rangers. The official motto they adopted was "Vigilans – the Watchers."

To the military brass, the results were astonishing.

The Rangers supplied vital intelligence about routes through terrain thought impassable, located water sources and potential camp sites and knew of ancient trails and the safe routes into shoal-choked bays and inlets.

Just as important, they provided a wealth of strategic information about logging roads and old skid trails which didn't appear on maps and were largely unknown to Canadian military intelligence – let alone Japanese.

"From the outset, it was therefore very obvious that the role of the P.C.M.R. from the point of view of operational intelligence was an extremely valuable one," an officer reported to Ottawa in 1943.

Today, the role of the modern Ranger hasn't changed much. There is technology that wasn't available 60 years ago, of course – much better radio equipment, the wonders of hand-held global positioning system receivers that can locate a position to within meters, lightweight and long-life field rations, microfiber rain gear, LED headlamps – but it still comes down to humping across brutal terrain under difficult conditions.

"We're the eyes and ears for the commander in Edmonton – and also the admiral (in Esquimalt)," Murray told me when we met in his cramped office. It's a room in what used to be a small hotel with a spectacular view of Victoria's skyline across the windswept straits outside Esquimalt harbour.

"We observe, report and assist but not in a tactical role, we are not expected to engage the enemy," he said.

I was there for a quick briefing before leaving on the 460 kilometer drive up Vancouver Island to Zeballos with Hryhoryshen, Cpl. Halay Stones and Pte. Leona Chaisson. Stones was a Rangers administration clerk out to get some field experience and Chaisson was an air force imaging technician – military talk for a photographer – who was there to record the exercise.

The 18-member patrol from Zeballos – the oldest is retired logger Dan Heppner, 65 this year – is so tightly knit that some, like Len Chester who moved away from the hamlet of 250 to take a job as a Vancouver paramedic, still commute to take part in exercises.

"I grew up here," he said. "I've been away for three years but I come back to do the (Ranger) exercises because I like it." On this exercise, Zeballos Patrol was joined by patrols from Gold River and Port Hardy.

The Rangers, wearing red uniform tee-shirts and hoodies adorned with the Ranger's insignia—a green shield with three maple leaves sharing one stem and a rifle crossed on a double-bitted axe – converged on the community hall to draw rations and hear the orders from patrol commander Steve Pederson.

But before that, Berlingette, Stones, Chaisson and the embedded columnist had to go through some qualifying safely training at the rifle range.

Some shot

I wish I could tell you otherwise, but *The Vancouver Sun*, which did okay with simple things like loading clips, clearing the breech and securing the safety, found itself in near eclipse when I came to actually hitting the target, whether from a prone, kneeling, standing or advancing position.

With targets the length of a football field from me, I could barely see the target, let alone hit the bull's-eye. I consistently shot high, raising dramatic puffs of dust as the bullets struck far away. I did discover a small cluster of holes in the outer ring of my last target, but I'm suspicious that Berlingette put them there out of sympathy.

That evening, the warrant officer conducted a weapon check, examining the bolts and looking in the breeches: "This one is shot out, we'll replace it later. The bolt's not right in this one – see how rough it is?"

Next came a refresher course in map and compass navigation: "What don't we use? True north. We use only grid north and magnetic north, map and compass," Hryhoryshen said. "We're working with a mixture of old and new maps here. Get your maps and your GPS in sync, I can't stress the importance of this enough."

The warrant officer provided instruction in how to calculate locations on the map grids to within 10 meters and offered a piece of advice – a sharp pencil, preferably a mechanical pencil with a fine lead, is as important a tool for bush navigation as a GPS. For one thing, it doesn't operate on batteries.

Make sure you have a really sharp pencil for making your grid marks," he told the patrol, "the thicker the pencil mark, the more you compound your margin of error."

Finally, Pederson delivered the orders. The Rangers were to board a flotilla of four small boats and travel by sea down Zeballos Inlet, then down Esperanza Inlet, then into the open ocean, turning southeast to the top end of Nootka Island. They would land at a remote site named Tongue Point and form into three sections.

Each section would carry one rifle for protection.

Up the Cone

The first section, under the command of Rick Kirkpatrick, the patrol commander from Port Hardy, would bushwhack across the island to a steep hill named Northwest Cone, climb it to reach an abandoned Second World War radar station and man it overnight, watching the coast with a portable radar unit.

The second, led by Bill Brooks, would cross he island, set up camp and conduct a surveillance patrol southeast along the shoreline, mapping trails and campsites while watching for a missing kayaker who was still the subject of an ongoing marine search. The third would establish a headquarters near the landing site.

Campfires were banned for the two patrolling units. Codes were set up – "No Duff" meant whatever followed was for real, not an exercise.

"Through me" meant a message was being relayed. "Sunray" would refer to the section commander, "Sunray Minor" to the second in command.

"Starlight" would be the medic. A radio watch was to be maintained at all times. Radio checks to be made every two hours.

At Tongue Point, still damp from the crossing, the boats rafted up and the Rangers went ashore on slippery, seaweed covered rocks, passing their heavy packs and gear quickly down a human chain. It was impressively swift considering the conditions. All three sections were ashore in 30 minutes and slogging across a long clam flat to set up the HQ that would coordinate communications and provide both reserve support and casualty response in case of an accident.

I left immediately with the first section, first taking an old plank road, then following the beached poles of a long-abandoned telegraph trail that dwindled away in the shore pines, then a narrow footpath through a bog and into the bug-infested underbrush.

I was grateful that all I was asked to carry was a long coil of rope and some extra water, not the heavy metal components of the radar unit. Still, I was pleased that at almost 60 I was able to keep pace with the younger soldiers.

Once again, the movement was swift, quiet and purposeful. Then we were at the foot of the Northwest Cone. It was steep, really steep. A party went to the top, stringing rope to ease the ascent for the others.

Roger Kervin, a tall, rangy [Ranger], swarmed up the squealing, corroded girders that were all that remained of the old tower, and lowered a rope to hoist the transmitter. The display unit remained on the ground.

Soon the Rangers were scanning the horizon from the rusty skeleton of a tower that had once watched for Japanese aircraft and fire balloons.

The balloons, loaded with incendiaries, were supposed to set B.C.'s forests ablaze, and an estimated 9,000 were launched against North America, carried across the Pacific by the jet stream. Precisely how many made it isn't known, but a woman and five children on a church outing were killed by one in Oregon. Others were located from the Northwest Territories to Wyoming.

The balloon attacks were kept quiet during the war for intelligence reasons but the Pacific Rangers, one government report later acknowledged, did more than any other combined operations to respond to the little-known threat.

There was no room for tents on their tiny summit of Northwest Cone and I left the Rangers in Section One cutting openings in the underbrush where they'd roll up in their tarps for the night. I went back down the rope and headed through the bush to headquarters.

I got back just before nightfall to discover the exercise overtaken by real life.

Clayton Brawn, whose grandfather was one of the original Rangers, was keeping the radio watch when at 7:50 p.m. he monitored a message from the Canadian Coast Guard at Tofino. A vessel was in distress. The six-metre craft had a flooded engine and was adrift about eight kilometers south-southwest of Catala Island in dangerous, reef-studded entrance to Nutchatlitz Inlet.

Hryhoryshen took charge, notified the surprised Coast Guard that a Canadian Rangers unit was in the immediate area, then broadcast the code word "No Duff" that informed the widely separated sections that the emergency was for real.

The topography made direct communications using VHF "line-of-site" radio difficult, so Hryhoryshen routed communications through the radar section on the top of the hill, which got to practice its "Through Me" procedures. Shortly after 9 pm. Hryhoryshen was able to notify the Coast Guard that one of the Ranger boats had the distressed vessel in tow.

That was cause for a pep talk from Hryhoryshen. "This is why we have Rangers," he told them. "This is why you are Rangers."

Up at first light

I didn't need a pep talk, myself, I needed some sleep after all that wet slogging through the bush and clambering up mountainsides, so I turned my soft, sore, urban body in early while others drank hot tea.

The next morning I was up at first light and off to find Section Two of the Ranger Patrol on the far side of the Island. That required a hike down a long inlet and then through dense underbrush. On the way, I passed through the hulk of a freighter. It had been wrecked on the outer coast, Chuck Syme of the Gold River Patrol told me. Then in some salvage operation, the hulk was cut into sections and skidded across the island. The scheme went broke and the ship was left in huge rusting pieces on the beach.

I found the surveillance section breaking camp where it had spent the night dossed down in dense salal just behind the windswept treeline.

It had patrolled the slippery, wind-blasted foreshore and was preparing to return to base camp, so I followed them single-file back through the seagrass as a heavy rain began to fall.

On the return, I spotted the tracks of a large, solitary wolf. They hadn't been there when I came by earlier that morning. They ran parallel to what must have been the high tide line, then turned sharply to the intersect the startled tracks of a big buck that suddenly veered off at right angles into the bush, the deep hoofprints indicating long bounds.

The whole ephemeral but eternal story of predator and prey was written on 100 meters of sand in shorthand more succinct than anything I've ever mastered.

So was the story of the Ranger patrol's exercise. The bootprints that came out of the bush and converged across the calm flats had already begun to fill with water and disappear as I arrived, Tail-end Charlie.

Once again the packs, considerably lighter now, were swiftly passed from hand to hand, the red-clad Watchers clambered back aboard the *Perseverance* and the eyes and ears of the generals returned to the world of jobs, mortgages and their kids' next report cards.

– 84 –

The Watchers: Unsung Arm of Military Work for Common Good of the Nation

Kenn Oliver

The Labradorian, 5 March 2007

They are your fathers and sons, mothers and daughters, brothers and sisters, coworkers, bosses, and friends. They are everyday people in communities across the country enlisted by the Canadian military to be their eyes and ears in areas where there is no military presence.

They are the Canadian Rangers.

And Labrador is proud to be home to 11 patrols and nine junior ranger patrols in communities from Nain down to Port Hope Simpson along the coast and inland patrols in Labrador City, Rigolet, Churchill Falls and a 28-member contingent in Happy Valley-Goose Bay.

Lieutenant Joe Anderson, a native of St. John's who has been living in Happy Valley-Goose Bay since 1958, has been a Ranger since the 1985 installment at 5 Wing during Operation Lightning Strike.

"It was a military exercise, one part of the forces acted as the enemy, another part wasn't. They had to see what they could capture and what they couldn't and they needed Rangers to take them around to places they couldn't because they didn't know the land," says the local patrol's commander.

Established in 1947, the Rangers are part-time reservists who provide a military presence in remote, isolated and coastal communities of Canada.

In addition and interestingly, they are the only DND members who are issued weapons to take home with them.

"We are issued a .303 rifle. If anything happens, they'll call you and say 'grab your weapon, we'll have the ammo, we need you here right away for the defense of the base or defense of the land," insists Lt. Anderson, a 21-year Ranger.

The Canadian Rangers website, at www.rangers.forces.gc.ca, says they are responsible for protecting Canada's sovereignty by reporting unusual activities or sightings, collecting local data of significance to the Canadian Forces, and conducting surveillance or sovereignty patrols as required.

The site describes Canadian Rangers as dedicated, knowledgeable members of the Forces who "play an important role in advancing public recognition of Canada's First Nations and Inuit Groups."

And it's the case in Labrador where many of the Rangers are aboriginal people, including 5 Wing patrol members Cpl. Sam Morris and Ranger Delbert Holwell, both originally of Cartwright.

Lt. Anderson says a people are particularly interested in joining the Goose Bay detachment because of its geographic location and interesting role within 5 Wing.

"I've got ex-patrol commanders from different patrols here, like Sam (Morris) was a patrol commander in Cartwright before he came here. I've got Rangers from all over the place," he boasts.

5 Wing Rangers

During the heyday of 5 Wing, the Rangers were a vital component in allied training. With their knowledge of the land and geography, they provided invaluable survival training, often in the harsh winter climate, to soldiers.

Mr. Anderson has been fortunate enough to work alongside British, Dutch, French, Italian, Germans and even the prestigious American Seals.

"They had their side of the camp, we had ours. The only way we could associate with them was when their commander came over and gave me their orders," recalls the former 60s and 70s American Forces mess hall kitchen mechanic.

Nowadays, under the new hotel concept at 5 Wing, the Rangers have been limited in their allied training. But over the last two months they have been providing arctic survival training to squadrons of the German Airborne Rangers (more akin to the American Rangers).

Ranger Holwell is experiencing his first go around with the Germans.

"(They) seem like they can well adapt to the Labrador conditions. From what I've seen they've done really well."

Cpl. Morris who has worked with other allies says the Germans "are as tough as any of them."

"Most of the military organizations are pretty tough, none of them really gripe, they realize it's part of their training," he contends.

The soldiers are learning skills that range from snowmobile driving to ice fishing to building lean-tos, which they are expected to spend a night in as part of their training.

Lt. Anderson says if they are deployed to cold weather climates, they'll be aptly prepared to survive thanks to the skills they learned from the his team.

One German soldier says the "experience and local knowledge is pretty impressive."

"Especially when it comes to lakes where you can drive, or where the ice is bad," he says.

While the training is there to serve the visiting German soldiers, it's also a benefit to the local Ranger patrols as it presents an opportunity for them to keep their skills sharp.

As of press time, the last group of German Airborne Rangers will have finished their arctic survival training at 5 Wing, leaving their newfound friends behind to continue their watch over the land and its people.

– 85 –

Op Tulugaq : Ranger Patrols from Nunavik and James Bay greet their Commanding Officer

Guylaine Fortin

***Servir*, March 2007**

From 17 to 22 January 2007, Major Guy Lang, Commanding Officer 2 Canadian Ranger Patrol Group (2CRPG), toured the Nunavik and James Bay areas to visit the communities where his patrols are located. The goal of this tour was to inspect the troops, share information with the Rangers, and meet with their respective village representatives.

Although he had visited the North several times since he took command in March 2006, Major Lang had not conducted a complete tour of all 17 villages where his Ranger patrols are located. Operation TULUGAQ, which means "raven" in Inuktitut, was just such an opportunity. By visiting three villages a day, he was able to meet with all the patrols.

Due to this heavy schedule, most of the meetings took place at airports. As landing helicopters created blizzard-like conditions, the Rangers stood proudly at attention, awaiting their CO. Often, while Major Lang proceeded with his inspection, you could see these men and women, their eyebrows covered in frost, shifting their feet slightly to keep warm. Once the review of the troops was over, everyone headed back to the airport to warm up.

Major Lang always stressed that he was pleased to meet with the patrol and thanked them for waiting for him and his team. "The Rangers are an important part of the Department of National Defence. As our eyes and ears here, you ensure a vital presence in this vast territory," he said.

A new way of training

Among the items on the agenda, there were announcements regarding a new training method and the purchase of new equipment. In order to improve training session efficiency, 2 CRPG will be implementing a two-phase training program. One year, patrols will receive training on search and rescue at sea; the following year, training will focus on land-based operations.

Another change is that patrols will train other patrols. "Your training will be more intensive and the scope of your duties will increase," Major Lang told his troops. "By having two or three villages work together, you will get to know one another and the weaker team members will learn from the stronger ones. If one day you are asked to work together, you will have common work methods and you will be more effective."

This change will also allow 2 CRPG to encourage its patrols to set up command posts from which sergeants and master corporals can direct future operations. Since Rangers are involved in all kinds of other activities, this could have positive repercussions for their communities.

New equipment

When the question of equipment arose, the Rangers were happy to learn that they would soon be issued new life jackets, anchors, two-way radios, or Mustang-type waterproof suits to better fulfill their mission. However, they were not pleased to hear that they would continue to use the .303-calibre rifle. Note that each Ranger is issued with this weapon and a supply of 300 bullets, which may be used for hunting.

"I know that some of you feel that these weapons are out of date. However, if we provided you with assault rifles, they would have to be kept in a weapons vault somewhere in the village. For hunters such as yourselves, this would not be very practical," explained Major Lang. "Even though your weapons may seem old, take good care of them. Clean them and keep them well-oiled, so that they are always fully functional. In fact, you should treat your weapon as well as you would treat a lover! One day, this weapon may save your life..."

If there is one piece of kit that the CO 2 CRPG would like to give to each of his Rangers, it is certainly a pair of warm boots. "I saw some of these men and women in running shoes, even though it was minus 35 outside," said Major Lang. "That's crazy! Even though the manual states that Ranger equipment should be light, these troops should be dressed according to the climate in which they operate." As a result, the Commanding Officer promised that, when he returned to Saint-Jean, he would do everything he could to find the funding to buy boots.

A valuable sharing of information

Even though the meetings were brief, they allowed the Rangers to raise a variety of issues, such at the training schedule, the pay deposit method and the consequences of losing a weapon. The mood was light as they discussed administrative headaches and presented their CO with a series of draft projects.

In a few villages, the Rangers asked about the possibility of deploying overseas or participating in UN missions. Major Lang reminded them that "The Rangers were not created to serve overseas. However, if you wish to do so, you can take part in these operations by enrolling in the Regular Force. Even though few Innu or Cree from Quebec have enrolled in the past, don't forget that during the Korean War, First Nation soldiers were recognized as being among our best warriors."

Before leaving each village, the CO 2 CRPG restated his confidence in his patrols: "I am proud of you and hope to have the opportunity to see you more often. Being a Ranger means being a soldier and a good citizen. It proves that you are courageous and ready to put your life on the line for your Nation. Be proud of your uniform and of what you do. I know that we can count on you."

Maj Lang will soon visit the Lower north shore.

– 86 –

Arctic 'Sucks In' Sovereignty Patrol: Ingenuity Sees Soldiers Press on Goa on Sea Ice and Land Canada Claims

Bob Weber

Edmonton Journal, 2 April 2007

Eureka, Nunavut—It didn't take long for the Arctic to take its toll on soldiers participating in an unprecedented snowmobile trek to enforce Canadian sovereignty.

A week ago, on the first day of their 5,000-kilometre patrol across and around Ellesmere Island, the 24 soldiers and Inuit Rangers lost two of their snowmobiles and all of their supply-laden komatiks, or sleds.

"We got sucked in," said Maj. Chris Bergeron, commander of the patrol, now pausing at the Eureka weather station on the way to [CFS] Alert at the island's northern tip.

The frozen river bed the patrol had been following had seemed to offer easy passage. But eventually the way was blocked by boulders bigger than their snowmobiles.

The travellers were forced to winch everything over the boulders --a process that tore the wooden sleds apart and punched holes in the transmissions of two machines, draining all their fluids.

"We knew we were going to have days like that," said Bergeron. "This is the North."

So ingenuity took over. First, patrollers lashed the sleds back together. Then they cut bits from some extra rubber snowmobile tracks, plugged up the holes in the transmission cases, topped up the fluids and headed out again.

It was one example of the determination required daily on Operation Nunalivut, an epic journey across land some of which hasn't been visited since the early Arctic explorers were there a century ago. The idea, said Bergeron, is to demonstrate to the world that the Canadian military is capable of operating on sea ice and land that Canada claims for its own.

Near the end of the patrol, the soldiers will place a metal Canadian flag on Ward Hunt Island, a rocky outcrop off the top of Ellesmere Island, which is used by explorers from all over the world as a jumping-off point for the North Pole.

"We have a special flag that will last for ages to make sure that people remember this is Canadian land," said Bergeron.

After the break in Eureka, the patrol will divide into three groups. One will head up the centre of Ellesmere. Another will swing west and follow the coast up to Alert, a trip that has never been done. A third, which includes a Mountie and fisheries officers, will cross to Alexandra Fiord on the east coast to check for any signs of Greenland Inuit crossing Nares Strait to hunt on Canadian land.

The days are gruelling. Patrols break out of their sleeping bags at 6 a.m. for breakfast. About 10 hours a day are spent on the trail. Half that is on the bucking backs of their snowmobiles. The rest of the time is spent either making repairs or getting past obstacles.

Blowing snow obscures landmarks so completely soldiers have been relying on GPS to navigate.

At day's end, they anchor their tents to their sleds so the wind doesn't blow them away. Then, hunkered down in the freezing dark, the Inuit hunters trade stories. One of the soldiers pulls out a harmonica and plays old Newfoundland tunes. Temperatures plummet to -40 C. Winds howl.

The patrol woke up one morning to find polar bear tracks near the tents.

"I do it for my community and my Canada," said Ranger Paul Ikuallaq of Gjoa Haven, Nunavut.

He also does it for his family. Ikuallaq is the grandson of famed Norwegian Arctic explorer Roald Amundsen, who fathered a child during a stay there.

"It's catching up with him," said Ikuallaq. The patrols are expected to reach Alert on April 10. "It's never been done before," said Bergeron. "It's history. I'm so proud to be here."

– 87 –

Where Only Rangers Tread

Kent Driscoll

Northern News Services, 9 April 2007

Resolute—When you are backing up the country's claim to the top of the world, bring the Canadian Rangers.

Operation Nunalivut has three Ranger patrols operating all over Ellesmere Island, and those patrols were packed with Nunavut members of the Rangers. If you are going to travel the High Arctic, it is wise to bring Nunavummiut.

The 24-member patrol is an all star team of local Ranger groups. Team leaders got to choose their own Rangers from the ranks, and to judge by the Inuktitut on most of the kamotiqs, Nunavut Rangers were well-represented.

Three Ranger patrols left Resolute on March 25. Rangers assembled their own kamotiqs from lumber the army flew to Resolute. They also picked up their military issue snowmobiles, Bombardier Scandic 500s.

The military loves their acronyms, and even something as ubiquitous as a snowmobile in Nunavut has a tag. The official military title for snowmobile is L.O.S.V. or "light over snow vehicle."

Two teams headed north to the Eureka weather station, while the third made its way east across Ellesmere Island to Alexandra Fiord.

The two teams in Eureka split up, with one patrolling the northwest shore and a second crossing the interior. Both teams will meet in Alert, on or around April 10.

Nunalivut is the name of the operation, and it means "land that is ours." Demonstrating Canadian control of that land is the major purpose of the exercise.

Patrol three will be looking for evidence of Greenlandic hunters on Canadian soil. They brought an RCMP officer with them, but arrests are not on the agenda. Cpl. Tom Cooke is along for the ride, but according to the Rangers commander, he left his handcuffs behind. Instead, he will deal with any trespassers in a very Canadian way.

"He said he would ask them over for tea, because it is cold," said Major Chris Bergeron, in command of the operation. "Then he would find out where they are from, and fill out a full report. The aim is not to catch people."

Patrol one will place a metal Canadian flag on Ward Hunt Island. Ward Hunt Island is the traditional starting point for North Pole-bound adventurers.

"We have all sorts of tourists there, and we have a special flag, which will last for ages. It says 'This is Canadian land,'" said Bergeron.

Rangers are making better time on every leg of the journey. On March 25, they were on the move for four hours and 20 minutes, and were stopped for four hours and 37 minutes.

By March 29, they were only resting for two hours and 57 minutes, and moving for seven hours.

On March 25, they only covered 144-kilometres, but on March 29 they travelled 242-kilometres.

"We are learning from our experience. If we don't make it, the next guy will, but we will make it," said Bergeron.

That first day was a test of the Rangers and their resolve. They made a mistake and ended up in a riverbed, which slowed down progress.

"We were sucked in by a riverbed. We destroyed two machines and most of the kamotiqs," said Bergeron.

Bergeron has obvious respect for the Rangers he works with.

"I'm so proud to be here. Without these guys, it would be impossible to do that," said Bergeron.

And he said communication between troops and Rangers is not a problem.

"The only language barrier is with me." Bergeron is a Francophone born in Trois-Rivieres, Que.

He also will not put his Rangers in danger. High winds delayed the patrols' departure from Eureka.

The media settled for a staged photo op, with the Rangers faking a departure for the cameras.

"When I've got wind like that, I won't put my men in a dangerous situation," said Bergeron.

– 88 –

Exercise Connects Cultures

Roxanna Thompson

Deh Cho Drum, 27 April 2007

Interesting and unexpected things often arise in the most unusual places.

On April 22 I was invited to visit the Fort Simpson airport to meet some of the soldiers involved in Operation Narwhal.

Driving out to the airport I was pretty sure I knew what to expect. Having written a few articles about the operation I knew all about troop numbers, the scenario, the types of aircraft that were in use and the history of the exercise. It was all information that I thought would serve me well.

Even with all my prep work, after being escorted to the maintenance garage at the airport where the soldiers are sleeping and eating, I was still amazed. This was still the Fort Simpson airport but it was filled with a different sense of purpose.

By treating the scenario as something that is happening in real life all the young soldiers were moving around with a sense of purpose. All the soldiers were in their uniforms and carrying a variety of interesting things like unloaded guns and radio equipment.

The surreal nature of the scene was added to by the Griffon helicopters, which definitely don't look like the helicopters normally seen in the area. They flew overhead continuously, slinging fuel from across the river.

And in the midst of the sea of green colored uniforms a few red sweaters stuck out like bright flags.

Although the soldiers and the helicopters were exciting, the most interesting part of the whole exercise was the interaction between the 10 local Canadian Rangers and the soldiers.

The Rangers were completely integrated into the scenario with three Rangers assigned to each of the three sections at the airport.

Although their primary duty was to provide protection from wildlife like bears, the Rangers have really been acting as ambassadors for the Deh Cho and the territory.

For many of the reservist soldiers from Newfoundland and Nova Scotia this is their first trip to the Northwest Territories. The Rangers have been showing them what life in the North is about.

The Rangers have shared a variety of local traditions with the soldiers including hunting with snares, bannock, some medicines, and how to collect dry wood. Some have even taught a few words of Slavey to the soldiers.

The soldiers I talked to all commented on how much they've been able to learn from the Rangers and how great they've been. Interestingly, the compliment also worked in reverse with many of the Rangers noting how great the soldiers have been and all the new things they've learned from them.

Cultural exchanges and increased levels of understanding can happen in the most unusual places, even during a Canadian Forces exercises.

– 89 –

Learning from the Canadian Rangers

Sgt Brad Phillips

The Maple Leaf, 13 June 2007

Norman Wells, Northwest Territories—When he began playing his instrument, the circle of soldiers surrounding Canadian Ranger Angus Shae of Fort Good Hope grew quiet and thoughtful. The haunting sounds flowed over the campsite and most in the audience, lost in thought, were clearly enjoying the performance.

Later, one of the braver soldiers in the audience asked the name of the instrument that he had entertained them with, a question that was on everyone's mind. A thoughtful pause and then, much to the delight of the other Rangers, the musician responded: "A flute."

The response was greeted by howls of laughter.

This misconception may have highlighted a difference between the Rangers and other soldiers taking part in Operation Narwhal 2007, but it also showed how enjoyable and important joint undertakings involving the Rangers and elements of the CF can be.

The mandate of the Canadian Rangers is to provide a military presence in those sparsely settled northern coastal and isolated areas of Canada that cannot conveniently or economically be provided for by other components of the CF.

During Op Narwhal 07, the Rangers were fully integrated with the land forces from the different Reserve brigades, right down to the section level. They acted as local experts of the area and subject matter experts on northern operations and living. Also, each Ranger brought along his trusty Lee Enfield in case any early rising predators, most likely bears, decided to join the fun and become part of the operation.

The Rangers were involved in all aspects of the operation and in all locations, including Norman Wells, Fort Simpson and Inuvik where all the major scenarios were played out. The Canadian Rangers present at Norman Wells came from either Fort Good Hope or Tulita, and for many it was their first experience working with members of the CF who were not part of the Canadian Rangers program.

Ranger Corporal Cyril Cardinal, part of an infantry section involved in vital point security, was getting used to pulling guard duty with the rest of the troops. "I think we're both learning from each other. Everybody is working pretty good together and seem to be having a lot of fun," he responded when asked how the operation was going and his involvement with the troops.

His comment seemed to reflect what all the Rangers, as well as the troops that had the opportunity to work side by side with the Rangers, were thinking.

For many of the soldiers, the highlight occurred near the end of the exercise when the Rangers had a chance to set up a tent in the hills near Norman Wells. At their campsite, the Rangers demonstrated how to set snares and gave tips for living off the land. They also prepared some of the local cuisine that the troops had the opportunity to appreciate. Some of these treats included: moose stew, a rack of caribou ribs, whitefish cooked beside the fire, bannock (homemade bread) and, of course, coffee and tea.

Cpl William Simmons of The Toronto Scottish Regiment summed up best the experience of working with the Rangers: "It was a great experience joining team with them."

– 90 –

Time is Right for Ranger Expansion

Northern News Services, 27 August 2007

Overshadowed in the federal government's Aug. 10 announcements of a deep-water port for Nanisivik and a military training centre for Resolute was word that the Canadian Rangers will expand nationwide to 5,000 members from 4,100.

In addition, they are to be provided with new gear, weapons and vehicles, an investment that will involve tens of millions of dollars over the next decade.

The Rangers routinely travel across remote areas of Canada, representing the military, carrying out search and rescue missions and leading soldiers on sovereignty patrols. Although they're capable of using modern technology such as global positioning systems (GPS), the Rangers possess skills that their Canadian Forces counterparts lack: familiarity with the vast expanse of land and ice and the ability to recognize weather conditions that can mean life or death.

The North is home to 1 Canadian Ranger Patrol Group. At about 1,500 members, it's the largest of five such regiments across Canada, with the others in Quebec, northern Ontario, Newfoundland and Labrador, the Prairies and Pacific west coast.

While satellites and radar have become primary sources of intelligence gathering for the world's leading militaries, they still require people on the ground. At a time when the High Arctic has become an international hotspot politically - for its natural resource riches and as a commercial shipping route - the Rangers are another means by which Canada can stake claim to disputed territory.

Not only are the Rangers a military presence, they are role models to youth in their communities. Easily identified by their signature red sweaters, these men and women play prominent roles at community events and are often a bridge between elders and youth in sharing knowledge of the land.

We will be better off with more of them.

– 91 –

Invisible Force in the North: Rangers guard sovereignty with old guns, radios

Don Martin

National Post, 26 October 2007

Ottawa—The Conservatives' Arctic reclamation project calls for unmanned aerial drones, ocean floor sensors, $3-billion worth of new patrol ships, a deep sea port and an expanded military base to drape the Maple Leaf across vast stretches of barren rock, ice and increasingly open water.

But our current guardians on the ground are a paramilitary force that carries 60-year-old rifles to fend off polar bears, provides its own snow, land or sea transportation, calls in news over personal satellite phones and exhausts holidays to stand on guard for us.

The Canadian Rangers seem to have been overlooked as the most experienced and economical way to reassert sovereignty claims over a resource-laden region with heightened potential as a shipping corridor in a globally warmed environment.

The Throne Speech gave them a backhanded salute -- pledging to increase their numbers by 20% while getting the name of the 4,100-strong quasi-military force wrong. Prime Minister Stephen Harper's address last week called them the Arctic Rangers, who don't actually exist.

But there's obvious potential to improve surveillance over a region claiming 75% of Canada's coastline using a force that's five times the size of our combat troop deployment in Kandahar yet costs less than the sticker price for three light-armoured vehicles.

Even though they're not whining about it, the Canadian Rangers have an obvious lag in equipment and status when stacked against regular soldiers.

Instead of a uniform, these recruits get a red sweatshirt, T-shirt, ball cap, vest, compass, first aid kit and, naturally, a toque. They're paid an armed forces reservists salary for the time they're on tour and must provide their own all-terrain vehicles, boats or snowmobiles to get around.

Baba Pedersen is the second family member in a three-generation string of Rangers from Kugluktuk, formerly known as Coppermine -- an accomplishment that earned the trio its very own postage stamp. (And just to get the obligatory conflict-of-interest declaration out of the way, Pedersen is related to me through marriage.)

But one could argue the 303 Enfields issued to every Ranger are a tad out of date. "My rifle is World War I vintage, but it works well in the cold and is good for shooting caribou," Pedersen says with a laugh. "If you were standing half a mile away, I could still drop you with it."

He has noticed, but isn't complaining, that Canadian soldiers have the best communications equipment money can buy while Rangers are given old high-frequency radios.

"When the military guys come on our patrols, they come equipped with all sorts of satellite and other technology. We use the stuff while they're here, but when they go, so does the technology," he says. "That's why I usually bring my own phone along."

Pedersen's tours, done during his vacation, take him on snowmobile dashes between the automated North Warning System outposts, calling in reports to North Bay's command centre. In the pursuit of re-asserted sovereignty, "we plant flags, take some pictures and move on," he says.

Pedersen has reported the odd sailboat that seemed out of place in the sea ice (no kidding) and called in unusual aircraft or weather balloons.

Perhaps the Rangers' most notorious apprehension was a Romanian who last year bravely boated to Grise Fjord on Ellesmere Island from Greenland, figuring he could mingle with the locals and eventually fly to Toronto unnoticed.

Unfortunately, he docked his boat 100 metres from a Ranger leader's home and didn't exactly blend into the local population, where he was only the second non-Inuit person in the community. He was detained by the Rangers and deported.

Nobody is quite sure what the Harper government has in mind for the Rangers beyond increasing their size by 900 members in the years to come. There's a leadership session in Yellowknife in a couple weeks where they might get fresh marching orders. Or not.

But given the positive impact an aggressive recruitment and training blitz would have on a region sagging under chronic social problems and high unemployment, an upgrade would seem to be a win-win move for both the locals and faraway Ottawa decision-makers.

"We always say something isn't ours unless we're standing on it," says Captain Conrad Schubert, spokesman for the Rangers program in Yellowknife, by way of explaining their role in protecting our sovereignty.

That makes this ragtag assortment of Inuit, who spend their vacations serving as our eyes and ears in a hostile but warming environment, Canada's best and brightest north stars.

– 92 –

Serving Together

Roxanna Thompson

Northern News Services, 22 November 2007

Sambaa K'e/Trout Lake—Like most couples Phoebe Punch and Clinton Betthale share many things, among them are matching red sweaters, the hallmark of the Canadian Rangers.

Betthale and Punch are a husband and wife Canadian Ranger duo from Trout Lake. Between them they have more than a decade of experience with the organization and hold the two highest ranks in their patrol.

Being a member of the Rangers gives a sense of being part of something larger, said Betthale.

"I'm part of the Trout Lake patrol and that means a lot to me," he said.

In the patrol Betthale has the second highest ranking, topped only by his wife. The difference in rank isn't a problem, said Punch.

"When I'm on duty I don't treat him as my husband. I treat him as another member of the patrol," she said.

Betthale said he doesn't have a problem taking orders from Punch because as soon as he puts on the red Ranger cap it's all part of the chain of command.

Sgt. Punch joined the Rangers in the late 1990s when a patrol was formed in Trout Lake. Signing up was a way to become the eyes and the ears of the community, she said.

Additional bonuses are the training exercises that, in Punch's opinion, are the best part of being a Ranger. During the exercises you get to go out on the land and there's always something new to learn, said Punch.

Being a Ranger also offers opportunities to be part of larger exercises. Punch was one of four Rangers from Trout Lake who took part in Operation Narwhal in Fort Simpson in April.

"That was a big learning experience," she said.

Training exercises and trips help keep members interested, said Punch. Over the years the Trout Lake patrol has grown to 15 members.

"The patrol is eager to go out on any training exercise outside of the community," Punch said.

Master Cpl. Clinton Betthale is among those who've joined the patrol.

After moving to Trout Lake in 2000, Betthale started talking with members of the patrol about their role. By 2003 he'd signed up.

"For me to be out on the land a lot more is what really caught my interest," said Betthale.

As a truck driver, Betthale said he'd watched a lot of land roll past his window but he'd never had the chance to live on it. Betthale said he's learned a lot of skills as a Ranger, like gun safety. He's also developed a sense of responsibility for the patrol and other Rangers.

"You aren't just one person. You're not there for yourself, you're part of the patrol," he said.

Rangers in Trout Lake will be practising taking orders during a training exercise in the community from Nov. 19 to 30. The event will involve a six-day trip on the land with an instructor from Yellowknife. Some new recruits will also be part of the event.

"They're very excited," said Punch.

After the exercise is finished a graduation parade will be held in the community along with a swearing-in ceremony for the new recruits. At that time the award for top shot will also be given to a member of the patrol.

Rangers are always anxious to find out who has won the award, said Betthale who has taken the honour before.

"It's just like waiting for your Christmas present," he said.

– 93 –

Rangers Arrive in Quebec City by Snowmobile!

Major Carlo De Ciccio

***Servir*, 13 February 2008**

On 28 January 2008, in a symbolic gesture, 37 Canadian Rangers and four instructors from 2 Canadian Ranger Patrol Group (2 CRPG), led by the Commander of Land Force Quebec Area and Joint Task Force East, Brigadier-General Christian Barabé, completed a historic journey on snowmobiles.

The participants had travelled from the province's three remote coastal regions (Nunavik, James Bay and the North Shore). Their first appearance was at Wendake, and the next day they stopped in front of the Hôtel de Ville in Quebec City. It had been a considerable challenge: a snowmobile expedition bringing together Rangers from 23 communities, including members of the Cree, Inuit, Montagnais and Naskapi First Nations, to salute the Huron-Wendat Nation and Quebec City residents as part of the festivities for Quebec City's 400th anniversary.

The origin of the project

The idea for the trip had been suggested by Brigadier-General Christian Barabé during a meeting with the Grand Chief of the Huron-Wendat Nation, Max *One-Onti* Gros-Louis, a little over a year earlier. "I wanted to not only set a challenge for our Rangers," General Barabé recalled, "but also highlight the presence and the importance of the First Nations in 2 CRPG." The Grand Chief of the Huron-Wendat Nation then suggested a meeting of First Nations peoples, just like the one that had been held 400 years earlier when Quebec City was founded by Samuel de Champlain.

Without further ado, the Commanding Officer of 2 CRPG, Major Guy Lang, was assigned the mission of organizing the expedition, with all of the challenges—anticipated and unanticipated—that it posed. "My objective, from the outset, was to do everything possible to ensure that my Rangers and my instructors could successfully complete the expedition," he said. "I'm very proud of them, because they all, without exception, finished the journey with no major incidents. They demonstrated that the Rangers of 2 CRPG—white, Cree, Inuit, Montagnais and Naskapi—could work as a team and that, together, we had the capability to successfully carry out tasks involving complex challenges."

First stop: Saguenay

The first objective for the three groups of Rangers was to rendezvous in Saguenay. The first group left Kuujjuak on 15 January with 22 Rangers from Inuit, Montagnais and Naskapi communities, under the command of Warrant Officer André Caron and Sergeant Philippe Rheault. The second group, made up of 6

Cree Rangers led by Warrant Officer Martin Prince, set out from Waskaganish (James Bay) on 20 January. The third group, comprising 9 Montagnais and white Rangers commanded by Warrant Officer François Duchesneau, departed Sept-Îles on 23 January.

The three groups arrived as planned in Saguenay on 24 and 25 January. To mark the occasion, four dignitaries—Jean Tremblay, Mayor of Saguenay; Lieutenant-General Andrew Leslie, Chief of the Land Staff; Brigadier-General Barabé; and Colonel Pierre Ruel, Commander of 3 Wing Bagotville—each rode a snowmobile to meet with the groups from James Bay and the North Shore on 25 January.

That evening, Lieutenant-General Leslie addressed the members of the expedition, who were all gathered in one place for the first time: "I am very proud that the Canadian Rangers are now under the responsibility of the Land Force. The expedition you are undertaking demonstrates the great capabilities of the Canadian Rangers. Your contribution and your expertise are important to the Canadian Forces. For all these reasons, we want to continue to increase our contribution to your training and the improvement of your equipment."

Wendake and Quebec City

After a journey of 2,600 kilometres for the Rangers from Nunavik and more than 1,500 kilometres for the groups from James Bay and the North Shore, the expedition finally arrived on 28 January at its first rendezvous in Wendake, the host nation for the celebration of Quebec City's 400th anniversary. Our Rangers were given a warm traditional welcome by the Huron-Wendat Nation and its Grand Chief. The chairman of the board of the Société du 400e anniversaire, Jean Leclerc, noted how proud he was of the CF's cooperation and involvement with the Huron-Wendat Nation in the anniversary activities.

The next day, 29 January, the members of the expedition completed the final metres of their journey by climbing the streets of the Old Town with their snowmobiles and their qamutiiks (dogsleds with two runners) as far as the Hôtel de Ville. "We made it!" the Rangers declared, seated on their snowmobiles in front of the city hall. During the reception hosted by the municipal government of Quebec City, a surprise awaited our Rangers: Bonhomme Carnaval! The Bonhomme was in great demand for photos with the guests. The expedition was the first big official activity held by the CF in 2008 to mark the 400th anniversary of the founding of Quebec City. It was very important that it be a success, since the CF will be holding a number of other events throughout the year to celebrate the occasion, as well as the presence and the heritage of the military in Quebec City throughout its 400-year history.

– 94 –

Rangers Reinforce Social Fabric in Canada's North: Diverse and unique group brings together families and northern communities

Cpl Jasper Schwartz

The Maple Leaf, 5 March 2008

Waskaganish, Quebec — Supporting Canada's northern communities and their families motivates individual Canadian Rangers.

When 37 Rangers from 2 Canadian Ranger Patrol Group (2 CRPG) recently completed an epic snowmobile trek to celebrate the 400th anniversary of Québec City, the first thing many of them did upon their arrival at the front steps of city hall was to call their families.

"That's who we do this for, our families and our communities. We are proud to be Rangers, and we hope our children will be proud of us," said Ranger Warrant Officer Henry Stewart, the senior Canadian Ranger in the James Bay Region.

Ranger Lindy Georgekish, a member of the Wemindji Patrol and a Cree from the James Bay area, completed over 1 800 km from January 20-29. He joined the Rangers two years ago looking both for a challenge and a chance to give back to his community.

"It feels good to be able to do something like this," he said. "I hope my son will be proud of seeing what I've done."

Norman "Junior" Cheezo, a Cree Ranger from the James Bay area, is following in his father's footsteps by joining the Rangers. He is hopeful his involvement will supplement his college studies in multimedia in the fall.

"It's a lot of fun," he said. "We don't get the chance to go out for days at a time anymore, live off the land, that kind of thing. We do it with the Ranger patrols. It's a whole different side of living up North."

More than 4 200 Canadian Rangers from 165 different communities work and live in the Canadian North. They are one of the most diverse and unique groups within the CF. The 700 Rangers of 2 CRPG, for example, consist of First Nations groups, i.e. Cree, Naskapi, Inuit, and Montagnais, as well as Anglophones and Francophones from the James Bay, Nunavik, and North shore regions.

Becoming a member of a Ranger patrol usually requires a commitment of at least 10 days a year to patrol the local wilderness. Members can also take CF courses on first aid, orienteering, rifle shooting and survival.

Canadian Rangers will play an ever-increasing role as the issue of arctic sovereignty takes on greater importance for Canada.

– 95 –

Arctic Shows its Benign and Beautiful Side in Epic Sovereignty Patrol

Bob Weber

Edmonton Sun, 7 April 2008

Ellesmere Island, Nunavut—Seven tundra wolves yawned and stretched, slowly rising from their snowbank bed near [CFS] Alert at the top of Ellesmere Island.

For a while, they watched us - a patrol of Canadian Rangers preparing for Operation Nunalivut, another epic trek across some of the world's most remote and rugged country in defence of Canada's Arctic sovereignty. Then they trotted off on their own wolfish business.

Soldiers at this most northerly post on earth are told not to feed or even approach the potentially dangerous pack, but this morning the magnificent animals seemed almost friendly as we readied our snow machines and packed our Inuit sleds, or komatiks.

I hoped that was some kind of sign.

I was joining the seven-member patrol for the first 220 kilometres of its journey across the top of the continent, from Alert to Ward Hunt Island in Quttinirpaaq National Park. The Arctic has sharp teeth and strong jaws, but it can be benign and beautiful and this was the side I wanted during the next five days.

Either way, I was with the right people.

The Rangers are a largely aboriginal group of reservists who are the military's northern eyes and ears, and those chosen for sovereignty patrols have exceptional land skills. Our group came from Trout Lake in the southern N.W.T. to Cambridge Bay along the Northwest Passage to Arctic Bay in the eastern High Arctic and ranged in age from young men in their 20s to elders.

"The guys that we've got are the best from their community," said Warrant Officer Dave Dunn, the rugged, hard-driving ex-paratrooper commanding the patrol. "These are the guys you want looking out for you on the land."

Samson Ejanqiaq, a 35-year-old grandfather, hunter and construction worker from Arctic Bay, Nunavut, would scout.

"He knows the ice, he knows the weather," said Dunn.

Adam Ukunqtunnaq, 27, a burly, affable Inuk from Gjoa Haven who liked to sing cheerful Inuktitut church songs, was our mechanic.

"If anything breaks down on the machine, all those Rangers will have that engine out, stripped apart, new piston in it and back on the road within an hour - guaranteed," Dunn promised. "In 40-below weather, with their hands bare."

Our "grandfather," as Samson called him, was David Issigaitok of Hall Beach, a smiling, unilingual elder valued for his long experience.

"He's a guy to listen to," Dunn said. "He'll be able to tell you by changes in the wind, moisture in the air, if there's a storm coming on."

On March 28, the eight of us roared off on snow machines towing three-metre komatiks loaded with hundreds of kilograms of rations, tents, spare parts, fuel, scientific equipment and even a generator. The sky was a calm, unclouded blue, the temperature in the -20s, and our view of Ellesmere's mountains and glaciers seemed to stretch forever in the crystalline air.

But it didn't take long for the Arctic to bare its fangs.

Within kilometres, we were caught in a vast boulder-field of rough sea ice. Samson and David did their best to pick through the car-sized ice chunks, but before long I had been bounced off my snow machine, got it stuck, rolled it twice and flipped my komatik a half-dozen times.

I ran with sweat under my heavy Arctic gear as we righted the machines and muscled them out of tight spots, and I wondered how long I could keep this up.

Finally, Sgt. Derek Dunn, a Ranger instructor who has led patrols across the Arctic, suggested that the shoreline might be smoother. Samson, guiding his machine and komatik through the rubble like a cowboy cutting cattle, found a path and, sure enough, a highway opened up - narrow, but blissfully smooth.

Ten hours we zoomed over the trail, pausing only for smoke breaks and lunch - soup, trail mix, caribou jerky and raw frozen muskox heart. Once, we stopped for an hour while Adam took apart an ignition right on the tundra to fix a stalling machine.

"There's glue all around those wires, eh?" he said. "It was all chipped from the cold and motion so I figured it was broke."

We paused again along the shoreline of a bay. Untouched mountains surrounded us, knuckles of ice stood aquamarine in the flooding sunlight and the Arctic stillness shouted silence.

Someone said, "It's beautiful," and that's all there was to say.

By 8 o'clock, we were snug in our doubled-walled tents, cocooned in heavy, army-issue sleeping bags. The ice beneath me was cold, but I slept anyway.

Good weather returned the next morning and we kept moving to Ward Hunt Island, our backs taking a pounding as we bounced over the hard-packed drifts. Although it seemed as if we were the first who ever burst across these seamless snows, we were anything but.

Royal Navy sailors travelled this route in 1875, towing sleds by hand. Robert Peary followed in 1906, and we visited the last point of land he saw before heading to the North Pole.

"(The route) is totally historical," said Ranger Doug Stern, a Parks Canada employee from Cambridge Bay. "Because we're limited so much by the Arctic ice pack and the mountainous terrain, we're following just that little bit of smooth ice at the bottom of the mountains."

Some predecessors remain mysterious.

We rode by a series of weathered old tents that Stern said were erected years ago by an Italian count. About 100 Greenland Inuit are said to have lived there once, but Stern said nobody knows what the count was up to.

By late afternoon, we reached our destination - a Parks Canada camp of insulated and heated tents on the north shore of Ward Hunt Island where we spent the next few days. The doorstep view was a white expanse of sea ice - a perfect canvas for the blues, pinks and oranges of a sun that never quite set but slid along a horizon so empty you could see the curvature of the earth.

We set into a routine of gathering scientific data, worthy work but a bit dull for Rangers eager to resume travelling.

"I came for the adventure," said Samson. "I want to shoot a muskox."

Samson would have to wait until we left the park, but animal signs were few anyway: ptarmigan, a week-old polar bear track and a small herd of muskox. We did, however, have other visitors.

One morning, four of us from two different tents reported hearing footsteps that night. Samson even got out of his sleeping bag to check, but found nothing. Nor were there any tracks the next morning.

"It was the Innugaguligajuk - the little people," said Samson.

The little people are said to stand about knee high and dress entirely in caribou skins. Nobody knows where they came from or what they do.

"If you leave them alone, they'll leave you alone," Samson said. "But if you attack them, they'll grow tall as a man right away. They're all over the Arctic."

The days passed taking measurements on the ice, and the evenings in hands of cribbage, bannock-baking and movie-watching on Derek's laptop. Eventually, the Twin Otter slated to take me back landed on the ice.

At the Eureka weather station, halfway down the Ellesmere coast, I towelled off from my first shower in a week, looked at the frostbite marks on my face and sifted through memories of five days in that icy vastness. I may never return, but my thoughts often will.

Outside, far in the distance, I could hear wolves howling.

– 96 –

Join the Rangers

Jennifer Hayduk

Terrace Standard, 21 May 2008

Canadian Rangers provide a lightly equipped and self-sufficient military presence in those areas of Canada that can't be conveniently or economically covered by other components of the Canadian Forces, and are best known for their work in the Arctic regions.

But not all Rangers are from the far north. Across Canada, 165 communities have Ranger Patrols, which employ about 4,000 people, most of whom already have full-time jobs but also find time for rangering.

There are two very active patrols in our area - one is based in Terrace and the other in Kitimat, and each is a small but important part of their community.

In June of 2007, you might have seen Rangers, in their red ball caps and sweaters bearing the green Ranger crest, along Queensway, on Brauns Island and in New Remo performing dike patrols and providing a security presence.

Rangers regularly take part in Remembrance Day parades, but the Rangers spend most of their time out of uniform and out of the public eye.

Ask most Rangers and they'll say one of the best things about being a Ranger is the training exercises, designed to develop and test bush craft and military skills.

These exercises focus on honing skills to prepare Rangers for real emergencies, and in the course of practising and learning, Rangers get to see and do some really exciting things.

In 2006, local Rangers, in conjunction with members of the Canadian Coast Guard, spent a few days in the ghost-town of Butedale, on Princess Royal Island southwest of Kitimat, where a Kermode bear and a couple of orcas made appearances within sight of the camp.

Members of the Terrace and Kitimat patrols participated in an exercise near Kemano, south of Kitimat, where 22 mountain goats provided entertainment for the watchers in the observation post set up near the camp and several grizzly bears in the area kept Rangers on two-hour shifts of bear watch/sentry duty through the night.

And most recently a large exercise near the little settlement of Doreen, east of Terrace, involving members from several patrols in Northwestern B.C., concluded with a raft trip down the Skeena River, through spectacular Kitselas Canyon.

Many smaller exercises, run in all seasons and weather conditions, lead Rangers in other places, all a little off the beaten track, and the coming months look like they'll be busy ones for the Terrace and Kitimat patrols.

Being a Ranger is not only about camping in really fantastic places. New recruits receive ten days of basic military training at CFB Esquimalt/ Camp Albert Head on Vancouver Island.

There the recruits are instructed in map, compass and GPS navigation, by both day and night, wilderness survival training, search and rescue techniques, marksmanship with the military-issue Lee Enfield .303 rifle and, because the Rangers are a branch of the Canadian Armed Forces, drill and marching, which can actually be quite fun.

Rangers also have periodic opportunities to attend RVs (Rendezvous), or training sessions, also held at Camp Albert Head, which provide a review of basic military skills, concentrating on adventure training.

The latest RV, in October 2007, included a day of sea-kayaking, day and night rappelling, more night navigation with a chance to use really cool night-vision goggles, a tour of the navy ship HMCS Brandon and a day on the shooting range with the .303 as well as the C7 automatic rifles like those used by the regular forces.

One highlight of the RV was an extensive and challenging obstacle/ confidence course of ropes, ladders and bridges designed to test strength, agility, balance and endurance.

The days can be long but the instructors are knowledgeable and competent, the barracks are comfortable and the food is excellent.

And if the lure of adventure and camaraderie, the chance to learn and the chance to lead are not reward enough for those dedicated to rangering, these part-time reservists get paid to do one of the things they like doing best - having a good time with new friends from all over Canada.

Terrace and Kitimat patrols are always looking for good people to join.

– 97 –

Rangers Seen as a Boon to Arctic Security: Reservists fly flag while serving as eyes and ears

Andrew Mayeda

National Post, 25 August 2008

Iqaluit—When Simigak Suvega and his men arrived here to showcase the skills of the Canadian Rangers, they did not bring their bolt-action rifles. They brought their knives.

Then, as a crowd of local Inuit residents and curious Canadian soldiers gathered around, they skinned four seals.

Their hands bathed in blood, the Rangers distributed the meat to residents, who carried away the entrails in plastic grocery bags or ate the flesh raw.

Some of the young soldiers, amazed by the spectacle, pulled out their cameras. But for many of the residents, it was a way of life that dates back millennia, and for the Rangers, it was just another day's work.

"That's one of the things that makes people join the Rangers, so they can help the communities," said Sergeant Suvega, a 60-year-old from Cape Dorset who has been a Ranger for 12 years.

Since taking power 2½ years ago, the government of Prime Minister Stephen Harper has promised to invest billions in military equipment and facilities to help Canada enforce its Arctic sovereignty, including navy patrol vessels and a new Coast Guard icebreaker.

But some Arctic experts say the government's best decision might have been the least flashy: the expansion of the Canadian Rangers by about 900 members.

"So far, the Harper government has revealed to us a sovereignty strategy that really amounts to a security agenda," said Whitney Lackenbauer, a Canadian International Council fellow and history professor at St. Jerome's University in Waterloo, Ont.

"It's investing in largely southern capabilities that can be deployed to the North. The Rangers represent something that's an investment in Northerners, for Northerners."

The Rangers trace their lineage to a volunteer militia established on the Pacific Coast during the Second World War to repel an invasion by the Japanese.

As the Cold War dawned, the Canadian government created the Rangers as a reserve force to protect Canada's remote communities -- a role they play to this day. In many cases, the Rangers are the only military presence in the isolated hamlets along Canada's Arctic coasts.

Technically, the Rangers are volunteer, part-time reservists, but they are unique from the rest of the Canadian Forces in many ways. The requirements for joining are not strict, and recruits receive only 10 days of basic training.

They are also lightly equipped. In addition to the Rangers' trademark red sweatshirt and red ball cap, they receive a .303 Lee Enfield rifle (a weapon that was standard issue during the First World War), an armband and a safety vest. Ranger patrols usually carry GPS devices, first-aid kits and shortwave radios.

On top of a small annual stipend, Rangers are paid only for official missions, but they can quit at any time. Or they could serve for years -- there is no retirement age.

But in an era of satellite surveillance and unmanned spy drones, the Rangers' presence has proven to be a low-tech, cost-effective way of asserting Canada's sovereignty.

"Those Rangers are the on-the-ground eyes and ears," said General Walt Natynczyk, Canada's chief of defence staff.

The Rangers are an under-appreciated but essential part of Canada's "stack" of surveillance systems, he said. "If they're out there on the ground [and] they see something suspicious, then they are the trigger. But if we actually see something from a satellite, from an aircraft, then they are the guys who are able to respond."

In addition to sovereignty and surveillance patrols, the Rangers are responsible for everything from inspecting NORAD radar stations to leading search-and-rescue missions.

During Operation Nanook, the $2.8-million Arctic sovereignty exercise that kicked off here last week, Rangers forces have been teaching traditional hunting, fishing and camping skills to Canadian soldiers from the south in the remote Baffin Island communities of Pangnirtung and Kimmirut.

"It's this environment that we need to get used to, and that's why the Rangers are very important to us, because they understand the environment. They know the climatic conditions. They know the terrain," said Brigadier-General David Millar, commander of Joint Task Force North.

The Rangers also offer a way for the military to build bridges with northern communities while encouraging them to safeguard their traditional skills and maintain independence, said Mr. Lackenbauer, who is writing a history of the force and has called the Rangers a "post-modern militia that works."

There are more than 4,500 Canadian Rangers across roughly 165 communities. In Nunavut, the vast majority of Rangers are Inuit, and many speak only Inuktitut.

– 98 –

At Home on the Land

Herb Mathisen

Northern News Services, 1 September 2008

Kimmirut/Lake Harbour—After spending more than five days on the land around Kimmirut with the Canadian Rangers, soldiers from units of the 33 Canadian Brigade Group had many humbling tales to tell and high praise for the group.

One of the most-talked-about stories was of a Canadian Ranger leading soldiers on patrol over hilly terrain.

The troops struggled to keep up with him and as they crested a hill, they expected to see him making his way down the other side, or just starting up the next hill.

They were shocked when they saw him sitting atop the next hill, off in the distance, smoking a cigarette while he waited for them to catch up.

"The troops called him 'Superman,'" said platoon commander Capt. Brian Lypps.

Superman, also known as Ranger Lysa Pitseolak, was modest about the tale.

"I was wearing shoes and they had on their boots," he said, adding the troops were also carrying their day gear.

"They were all sweaty," he laughed.

In total, 31 soldiers were in Kimmirut with others deployed around Pangnirtung for presence and point patrols as part of Operation Nanook - a Canadian Forces sovereignty exercise.

For nearly six days, Canadian Rangers trained the soldiers to survive off the land.

"They have to know how to survive in the cold weather and good weather like this," said Canadian Ranger Louie Qimirpik on a sunny Saturday morning.

Rangers showed soldiers where to set up camps, how to hunt small and big game and prepare their catch and what water was safe to drink.

They brought soldiers out to fishing spots to gather mussels and pick berries.

Sgt. James Vogl, who recently completed a tour of Afghanistan, was grateful for the experience with the Rangers, who familiarized them in an environment where Vogl admitted troops were "out of their element."

"The Rangers have been helpful, showing us how to survive off the land above the treeline," said Vogl.

Asked whether he thought there would be an increased military presence in the Arctic in the future, Vogl responded: "There is always a presence here with the Rangers."

Vogl has trained with the Rangers once before, in the winter, in Resolute.

"They took care of us up there like they did here," he said.

Lypps said Rangers were the only ones on patrol with ammunition, providing protection to troops from polar bears.

Vogl said the Rangers also helped soldiers understand Inuit culture.

Soldiers and Rangers pulled plenty of cod out of the falls by Soper Lake on Saturday.

Qimirpik and Appa Josephie demonstrated how to gut and fillet the cod, which was then shared with members of the community.

"Once we get food, we pass it on to the community for people who don't have hunting equipment," said Qimirpik. "That's pretty much the tradition for our community, Kimmirut, and we don't want to lose it."

Rangers initiated the troops by getting them to taste the cod's raw liver.

Aside from a few rolled ankles and the inevitable blisters, soldiers made it through the training intact with a strong camaraderie built between the two groups.

Canadian Ranger Sgt. Jamesie Kootoo said he hoped there would be more winter training.

"We like to do that in the winter time," said Kootoo, adding that winter conditions provided more survival training exercise than the summer, including iglu and snow cave building and ice-fishing.

"It's harder," said Kootoo.

Kootoo is also concerned with the significant drop in the number of Canadian Rangers in Kimmirut.

"There are 12," he said. "I used to have more than 30."

Qimirpik said government red tape has been an issue.

Rangers in Kimmirut sometimes have to wait two months for paycheques, which travel from Ottawa to Yellowknife to Iqaluit before making it into the community.

"Sometimes Rangers don't like to wait a couple months to get paid," said Qimirpik.

Defence minister Peter MacKay recently announced $3.6 million for the Junior Rangers over five years.

Kootoo said he is encouraged by the increase in the number of Junior Rangers in the community, saying it's more than 20 now.

He added any new money was good news for the program.

Qimirpik's son is involved with the Junior Rangers. He said he believes the program is valuable not only in keeping the Canadian Rangers strong, but also to preserve traditional knowledge on the land.

"That's a good thing to pass on knowledge to our Junior Rangers, because they're our future," said Qimirpik.

Lypps said he was proud of all the tasks accomplished by the group, while they were able to build strong friendships.

"I think we've developed a mutual respect for one another and the skills we have," he said. "We're all Canadians in the end."

– 99 –

Rangers to Increase in Numbers and Patrols

Sgt Peter Moon

3 CRPG *Newsletter*, February 2009

Major Guy Ingram says 2009 will be "a very busy year for the Canadian Rangers and the Junior Canadian Rangers."

The number of Canadian Ranger patrols across Northern Ontario will increase from 15 to 21 over the next three years.

"I've been mandated to open six new patrols in Northern Ontario in the next three years," said Major Guy Ingram, commanding officer of 3rd Canadian Ranger Patrol Group. "I'm in the process of visiting the 17 communities that have asked for Ranger patrols. I've visited three so far. We expect to open a new patrol by February or March."

The increase in the number of patrols will see significant increases in the number of Canadian Rangers across Ontario's Far North. There are currently about 400.

"This year will be a big year for expansion for the Canadian Rangers across Canada," Major Ingram said. "There are currently 4,200 Canadian Rangers in 165 remote and isolated communities and that is going to increase.

"The army has agreed to issue a range of new personal kit for Rangers. That's 10 items of personal kit - all clothing - and it will be issued this year. It could be a parka, pants, boots, a Gerber knife, mitts, underwear, stuff like that. We will also be seeing a lot more equipment being put into our communities to support the Rangers."

The increases in numbers and equipment follow on a busy year for the Canadian Rangers of Northern Ontario. "Last year was probably one of the best years we've had," Major Ingram said. "Over the last 10 years I've seen 3CRPG grow from a minor unit in the eyes of the military into what I consider to be an operational unit with a record of major successes. We've punched far above our weight for a unit of our size."

The pace of operations, training and other activities by Ontario's Rangers in support of military activities in the province has placed a strain, he said, on the 24 full-time soldiers who staff the unit's headquarters at Canadian Forces Base Borden, near Barrie. "They are all busy all of the time," he said. "We need 40 people to do the job at the rate we are going now."

The level of training reached by the Canadian Rangers in 2008 was "phenomenal," he said, but Rangers can expect to see more challenging training this year to provide them with increased skills for domestic emergencies.

"We've got to come up with better training methods to engage them and make training interesting for them. Instead of just sitting in a classroom to learn first aid, for example, they will have to go out into the bush afterwards and deal with simulated casualties, putting into practice what they have learned in class. That combination of training will help them save lives in the future."

There are about 500 Junior Canadian Rangers in 15 patrols in Northern Ontario and their numbers will grow as the number of Canadian Ranger patrols increase. New clothing, adapted from the army cadet programme, will see Junior Rangers receiving new jackets, toques, hats, pants, hiking boots and other equipment this year.

"We had 76 Junior Canadian Rangers at Camp Loon, our annual summer camp near Geraldton, last year," Major Ingram said. "We are going to double that number this year by operating the camp over a longer period of time.

"I tell you, 2009 is going to be a busy year for the Canadian Rangers and the Junior Canadian Rangers. It's going to be a very good year."

– 100 –

Rangers Provide Wilderness Training for Canadian Soldiers in Timmins

Len Gillis

Timmins Times, 4 March 2009

Forty-one year old Joe Lazarus doesn't know a lot about Canada's sovereignty concerns in the Arctic, but the Kashechewan native takes pride in teaching Canadian soldiers how to live and survive in the extremes of a Canadian winter.

Lazarus is one of a group of seven Canadian Rangers that was in Timmins for more than a week, teaching wilderness skills to regular and reserve members of the Canadian Forces.

The Canadian Rangers are comprised of aboriginal men and women who serve in a special branch of the Canadian Forces in Canada's remote northern regions.

The exercise held in Timmins was directed at ensuring Canadian Forces troops can survive and operate in the far north, said Major Guy Ingram, the commanding officer of the 3rd Canadian Ranger Patrol Group from Canadian Forces Base Borden.

"We've got a bunch of southern soldiers who aren't used to the north anymore. That's basically what it comes down to," said Ingram.

"The Canadian Forces are involved in a number of things ... As you know the government of Canada has said we shall be involved in the Arctic. And that's why we're striving to make sure the soldiers are capable of operating there," said Ingram.

The six men and one woman, all from Kashechewan, set up camp west of the city and soldiers were brought in from Sudbury, North Bay and Sault Ste. Marie for the training.

"For this exercise we've got members of the Algonquin Regiment, the 26th Service Battalion and the 2nd Irish Regiment of Canada out of Sudbury, so it's basically all Northern soldiers, a mix of infantry and combat service support," said Capt. Matt Baxter of the Algonquin Regiment, A-company from North Bay.

The men and women reservists were taught such things as how to dress warmly and stay active, how to survive falling through the ice of a lake, how to light a fire quickly and how to build shelters out of piles of snow and sticks.

Building shelters is the specialty of Master Corporal Joe Lazarus. As small groups of soldiers watched and listened, Lazarus outlined how to build impromptu structures as quinzies and hand-made shelters designed to keep you warm and alive.

"This is what my dad taught me," said Lazarus. "I thought it would never take me anywhere, but I am very proud of what I learned and what I teach," he said.

Lazarus doesn't speak of survival lightly. He knows what he's talking about.

"One time it took me 33 hours to get home," Lazarus recalls of a snowstorm on the James Bay coast. "It was very awful. It was on a fishing trip." Lazarus had only his winter clothing and snowshoes. To make things worse, he hadn't caught any fish. But he survived and three days later he was home.

"It can be very hard and very brutal out there in the winter to maintain yourself," he said.

"The only thing that pops into your head like when you're stranded is death. Am I gonna die here. That's what pops into your mind, even though you don't want to think about it. It's always gonna be there. But you have to maintain yourself, focus on yourself and realize where you are," said Lazarus.

It gives the soldiers pause as they look at him and take in what Lazarus tells them.

Capt. Baxter says the Rangers have the kind of experience that soldiers cannot ignore.

"It's very applicable to being a soldier because we have to survive out in the bush and we may not have the resources required to do a mission or we may run into trouble and learning these skills helps us to survive, so its great training," he said.

Major Ingram agrees and says the First Nations people who make up the Rangers are "probably the last people who have that expertise and knowledge and for them to share that with the soldiers is very important to us.

"I don't know about other Northern countries," said Ingram. "I think the Rangers themselves are unique to the world. I know that for a fact. There are no other Rangers in the world that pass this on, but I'm sure there's countries like Russia and Finland and they are capable of operating in cold environments as well. So this is good and every year it gets better."

– 101 –

Battling the Most Rugged Terrain

Cheryl MacLeod

The Maple Leaf, 11 March 2009

Points North Landing, Sask. — After many hours of driving through the snow and extreme cold on trails only the strong would travel, the road ahead leads to a small community and, hopefully, rest for the night. To the travellers' surprise, there are people cheering, waving and acknowledging what they are doing.

This is something the Rangers of the 4th Canadian Rangers Patrol Group (4 CRPG) encounter as they move through small northern communities during Exercise Western Spirit. The exercise has Canadian Rangers and Army reservists of 4 CRPG—with support from 440 Transport Squadron and HMCS Calgary—crossing all four western Canadian provinces by snowmobile with minimal support over 30 days, covering roughly 3 400 km of Canada's most rugged terrain.

More than a year in planning, the purpose of Ex Western Spirit is twofold, says 4 CRPG CO Major Tim Byers. "Most important is to challenge us [Rangers] and increase our skills sets," he says. "Secondly, and equally as important, is to create a better awareness of the Canadian Rangers, especially below the 60th [parallel]. Most Canadians—and even the military—when they hear Canadian Rangers, they think Nunavut and the Yukon, and that's not true. There are more Rangers below the 60th than above."

One of Maj Byers' biggest challenges has been the logistics of bringing Rangers from four different provinces together in one spot, getting them kitted out and then organizing the flights and HMCS Calgary. They started out in British Columbia. "On the trails, our biggest challenge hasn't been the weather, but not having snow," Maj Byers says with a smile. "The lack of snow at one point in B.C. was a surprise. It literally looked like April; it was just freaky for this time of year."

Adjustments were made and the exercise continued without much delay. The Rangers are the lead in this plan, guiding the team through every leg of the exercise, repairing the machines and using their skills and instinct to keep things on schedule. "These things are second nature to them, because this is how they live," says Maj Byers. "The amount of the skill they perform with in their environment is outstanding. You cannot take a Reg[ular] Force unit, put them in this environment and have them perform better than they [Rangers] do."

Many of their nightly stops are in very remote areas, which means sleeping in tents and cooking on portable stoves in frigid temperatures. But this doesn't seem to bother the more than 100 Rangers, in total, participating in the various legs of the trip. "What an opportunity to see Canada!" says Ranger Maryanne Wettlaufer, from the Port Clements, B.C., Ranger Patrol.

Rgr Wettlaufer is an artist, and sees the exercise as a great opportunity not only for inspiration for her landscape and cultural sketches, but also to get more involved with the Ranger program and help document the exercise. Her biggest challenge has been not having the solitude and freedom to be on her own to sit and sketch. "I think there is also a pro to that," she says. "You're with the group and you really get to know and gel with these people, and feel like you are part of a team."

Rgr Wettlaufer believes that, once she has some down time, she will feel the impact this adventure has had on her as a person. "It will become much clearer, what this experience has done for me," she says, "not only as a person but as an artist. Every day the light changes; every day the landscape changes. This has been a great experience."

The biggest challenge for Rgr Camille Hamilton, a mechanic, is keeping the snowmobiles moving on the trail. "Things went well," he says. "No major problems."

Living in Churchill, Man., Rgr Hamilton sees very few trees. "These big trees are a surprise." He believes Rangers play an important role in Northern sovereignty for many reasons, but mostly because "we know the land so well," he says. "It's a way of life for us."

For Rgr Cherie Nickel, from 100 Mile House, B.C., preparing for her 42 days away from home was the biggest challenge. "The fact is, we have to carry everything we need," she says. "My challenge was having to pack light." Tom, her husband of 18 years who's also part of the exercise, says, "she's never had to pack this light."

What the couple has missed most since leaving home in mid-January is their daughter, who is being looked after by her grandparents. "We also miss the mountains, and our own bed," says Cherie. "It's hard to get a good night's sleep with all the snoring."

– 102 –

Northern General Heaps Praise on Nunavut Rangers

Chris Windeyer

Nunatsiaq News, 13 March 2009

"People always ask me how I protect sovereignty. My answer is always one simple word: Rangers"

Gjoa Haven—Canada's top northern soldier toured six Nunavut communities this past week to say thank-you to the Canadian Rangers he says are key to protecting Canada's Arctic sovereignty.

Brig. Gen. David Millar stopped in eight Nunavut communities over five days starting Feb. 28 in Rankin Inlet and ending March 4 in Gjoa Haven. There are Ranger patrols in 56 communities and Millar said he wants to visit them all.

"[This trip is to] get me out on the land where our Rangers operate, meet with them, understand their concerns, but also to motivate and... convey to them how important their role is in defending and representing sovereignty in the North," Millar said during some rare down time at the E.J. Lyall community centre in Taloyoak.

In Qikiqtarjuaq, Millar inspected the local company of Rangers and Junior Rangers, telling them they are on the front line defending Canadian sovereignty in the Arctic.

"People always ask me how I protect sovereignty. My answer is always one simple word: Rangers," Millar told each of the patrols.

Sgt. Stevie Audlaqiaq has been a Ranger for 37 years, ever since the Canadian Forces first asked him to show their soldiers how to survive on the land during exercises at the old Cape Dyer DEW Line site.

Audlaqiaq said he and other Rangers from Qikiqtarjuaq still go out to Cape Dyer, and other sites that are now part of the North Warning System, to check for damage and oil leaks, and to clear snow when necessary. That happens three times a year.

The rest of the time Rangers practice their marksmanship, take part in community events and help with search and rescue operations.

In his remarks, Millar also praised the Junior Rangers and said he hoped they would grow up to be full-fledged Rangers. Audlaqiaq, whose own daughter and granddaughter went through the program, said Junior Rangers learn discipline, which helps them keep out of trouble.

"This is pretty helpful for them," Audlaqiaq said.

Millar's visit also serves as a chance to see if local patrols need new equipment. Rangers in Pond Inlet recently got new boots, but the treads are slippery and Sgt. Norman Simonie asked Millar for rubber cramp-ons. The general said he'd have some sent out.

"As soon as we get back to Yellowknife I'll ship out 37 [pairs] to them," Millar said later. And if Rangers in Pond on are having that problem with the new boots, other patrols likely are too. It's another reason why Millar thinks it's important to get out on the road.

Simonie said he was grateful for the chance to meet with Millar and was honoured by his visit.

"A general that's taking care of us coming into our community makes it very special," he said.

For Pvt. Jacopie Maktar, also of Pond Inlet, Millar's visit was an honour for the whole family. Maktar, his brothers Joseph Maktar, Bernard Maktar and Peter Oolateeta, and brother-in-law George Satuqsi, are all members of the Pond Inlet patrol.

Jacopie said he joined the Rangers eight years ago because his father joined the militia reserves back in 1953.

"I would like to keep his tradition," Maktar said.

In Clyde River, Millar told the local Ranger patrol their Sgt. Levi Palituq "is famous in the North" for being the only Ranger to be awarded the medal of bravery for trying to save a drowning man.

Seeing Millar off at the airport, Palituq told the general he was honoured by the visit.

"I can honestly say that I'm a proud Canadian and I love this," Palituq said.

Canadian Rangers are the leading edge of Canada's efforts to assert its sovereignty in the North.

They travel and hunt on lands that may one day play into the geopolitical designs of governments and military planners in Yellowknife, Ottawa, Washington, Copenhagen, or Moscow.

Millar said Ranger patrols in all of those 56 northern communities need to be acknowledged for the work they do on Canada's behalf.

"They are proud Canadians and representing Canada, but no one is saying thank you."

– 103 –

Canadian Rangers Go Back to the Basics

Lieut Shalako Smith

The Labradorian, 20 July 2009

Upon first seeing this grey-haired group of 40 and 50 something Canadian Rangers, each adorned in the Canadian Ranger uniform of a red hoodie with the Ranger crest, green combat pants and red ball cap, one might be surprised to know that they are the newest group of candidates for Ranger basic training.

For this course you won't see red-faced, CADPAT wearing instructors barking orders or row upon row of breathless candidates pumping out push-ups - things are a little more relaxed on this basic training course.

Nearly 30 candidates from Cartwright, Bonavista, Cape Freels as well as many other communities across Labrador including Happy Valley-Goose Bay are currently at 5 Wing for a pilot course for Development Period (DP) 1 "affectionately" referred to as basic training.

This is the first time this pilot course will be offered to members of 5 Canadian Ranger Patrol Group from Newfoundland and Labrador and it is open, not only to new members but also to currently serving Canadian Rangers.

An instructor for the two-week Canadian Ranger DP 1 course is Sgt. Anthony Dubourdieu. He says, for the Rangers that have been serving for a while this course will serve as a good refresher of Canadian Ranger traditions and skills, CF knowledge, and field craft.

He laughs a little when he thinks of the differences between this Ranger basic training and general basic training new military recruits go through.

"Oh, this is much different," says Sgt. Dubourdieu "[the staff] is able to be far friendlier on a course like this"

A valid reason for the lighter demeanor is the difference in age group. "We're not dealing with kids here. These are all responsible adults" he said.

While some of the standards for this course may be a little bit different from a regular basic training; with candidate not being expected to shave their faces or complete a physical fitness test; there are many similarities.

"This course will still offer lectures and practical training for basic drill, basic service knowledge [of the military], and personal weapons handling and safety," said. Sgt. Dubourdieu.

Being that it is a course for Canadian Rangers they will also learn Ranger-specific history, knowledge and skills as well as Search and Rescue (SAR) skills.

The Commander of 5 Wing, LCol. Brian Bowerman is happy that 5 Wing is hosting the Canadian Rangers.

“The Rangers are a valued part of the Canadian Forces Defence Team,” he said, “they help complete a robust and well-rounded Search and Rescue package employable by the CF in service of Canadians”.

“I’m certain that the skills and training that they are receiving here in 5 Wing will help make them valued members of their local community’s patrol,” he said.

The DP 1 course started on 7 July and will last until 21 July. A DP 2 course, which will reinforce some of the skills of this course and focus on teaching leadership skills, will start at the end of the month through the middle of August.

– 104 –

Rangers Erect Grave Markers

Sgt Peter Moon

Wawatay News, 12 November 2009

The formerly unmarked graves of three Canadian Rangers in two communities in northern Ontario now have permanent military grave markers, thanks to the Last Post Fund. They are the first grave sites of any Canadian Rangers across Canada to receive them under a new grave stone program that recognizes the military service of Canadian Rangers after they die.

The grave markers were dedicated in October at special ceremonies attended by the Rangers' families, the chiefs of their First Nations, students and community members, and Canadian Rangers and Junior Canadian Rangers. All three markers bear the image of a flying eagle.

"There was nothing on my brother's grave, just flowers," said Madeleine Nakogee of Fort Albany as she stood next to the newly erected marker for Ranger Michael Hookimaw, her brother. "There was a wooden plaque we made for him when he died but it rotted away. So this stone is nice. We are all very proud of it. It's a nice feeling to see him recognized like this."

The Last Post Fund delivers funeral and burial programs for Veterans Affairs Canada. It recently amended its special grave marker program, which provides military grave markers for veteran's graves if they have not had a permanent market for five years or more, to include not only veterans of the South African War, First and Second World Wars, and the Korean Conflict but former members of the Canadian Forces who have passed basic training and been honourably discharged. Canadian Rangers are members of the Canadian Forces.

"This shows once again that we continue to honour our veterans," said Major Guy Ingram, commanding officer of 3rd Canadian Ranger Patrol Group. "The Last Post Fund has decided to help honour the Canadian Rangers and we've embarked on a mission to ensure that every Canadian Ranger and First or Second World War veteran or Korean Conflict veteran in the Far North of Ontario is properly identified in their gravesite by a military headstone."

In addition to the ceremony in Fort Albany a second ceremony was held in Sandy Lake, where two markers were dedicated. A feature of the dedication, which was conducted by a military chaplain, was the involvement of Ranger Peter Fiddler, a local elder, who said a prayer and smudged the head stones.

One of the two markers in Sandy Lake was for the unmarked grave of Ranger Delilah Meekis. "When she passed away we wrote this song about her flying away," said her daughter, Donna Brunton Meekis. "So this eagle on the stone is very appropriate."

The other marker was for Ranger Frederick Meekis. "All the family is very thrilled by this," said his son, Ranger Menashi Meekis. "We are very happy with the design of the stone. The eagle is very appropriate."

Air Creebec provided free air transportion for the grave marker in Fort Albany and Wasaya Airways provided free transportation for the two markers that were flown to Sandy Lake.

– 105 –

Rangers Battle Surprise Blizzard in Successful Search

Sgt Peter Moon

3 CRPG *Newsletter*, February 2010

An unexpected blizzard forced Canadian Rangers to head out on a spring search and rescue mission in the dark and in near-zero visibility conditions.

Sergeant Bellamie Bighead of the Canadian Ranger patrol in Kingfisher Lake said the Rangers were asked to respond when a sudden change in weather left five people stranded on the land.

The emergency began around 7 p.m. on March 24 as temperatures plunged within a five-hour period from –7C to –23C. Heavy snow and sudden winds gusting to 50 kph caused near white-out conditions.

While other Rangers assembled food, clothing and fuel supplies and Ranger Bradley McKay manned a communications command post, Sergeant Bighead, Master Corporal Ronnie Sainnawap and Ranger Obadiah Winter, left the community on snowmobiles for a fishing site on Mariah Lake, where a 68-year-old diabetic man, who had gone to check his nets, was reported to be stranded with a second man and his nine-year-old great-nephew. Because of the mild weather, they had left Kingfisher Lake only wearing fleece jackets and caps and without food. They were only able to communicate intermittently with the community by radio phone. There were grave concerns for the older man's health.

"We drove through the blizzard," Sergeant Bighead said. "It was quite an experience. It was like driving to a place you can't see, basically. We were going very slowly and cautiously. It was hard going and very cold."

When they reached the fishing site they found the diabetic man missing and the second man and his nephew huddled around a fire. The three had tried to snowmobile back to Kingfisher Lake but they got separated in the blizzard. The diabetic man eventually managed to reach the community. The second man and the boy returned to the fishing site because they knew that that was where searchers would look for them.

The Rangers radioed for food, clothing and fuel. Master Corporal Andrew Mamakwa, Ranger Luke Winter, and Edward Sugarhead, a community member, arrived with the supplies shortly after midnight. The man and boy were given warm clothing and food. "They appreciated it," Sergeant Bellamie said.

The Kingfisher Lake patrol was formed last January and without it, Sergeant Bellamie said, the outcome of the emergency might have been very different.

"It could have been disastrous," he said. "The army training helped us a lot. If we hadn't had a Ranger patrol I wouldn't have known who to call for help in organizing the search and rescue. All the Rangers I called and asked for help said yes, right away, every one of them. Without the Ranger patrol I might have been out there doing a search on my own. That's the difference a Ranger patrol can make. Everyone knows how to work together."

"I'm ecstatic about how they conducted themselves," said Chief James Mamawka. "They did a tremendous job. We've wanted the Rangers for a long time and now we've got them. It makes people feel good."

– 106 –

Britons' Trip an 'All-Round Super Experience'

Dan Davidson

Whitehorse Star, 19 March 2010

Dawson City—Arctic Roller 2010 arrived back in Dawson City late on March 8, after a somewhat harrowing 780-km drive from Inuvik where they had spent the night after leaving Tuktoyaktuk.

They had arrived in Dawson on Feb. 23 at the end of a 3,000-km journey from the British Army Training Unit Suffield, located within Canadian Forces Base Suffield in Alberta.

British armored groups have been training in Alberta on and off since the Second World War, and BATUS was formally established in 1972.

While the Alberta prairie originally doubled for exercises in which the former Soviet Union was the battleground, these days it is used to prepare the troops for exercises in Afghanistan.

The base has around 200 permanent staff on two year postings. During the year, other people rotate, with as many as 1,500 to 2,000 additional troops brought in for training, which runs from May to October.

As Warrant Officer Tracy Hargreaves explained in an interview on March 9, the aim of the Arctic Roller exercises, which have been going on now for about 15 years, is to conduct survival training with the Canadian Rangers near Dawson City and Tuktoyaktuk. Twelve members of BATUS got to make this trip, which is high on the list of things they like to do.

"We bid for places, and if we're lucky enough, we get selected."

From Feb. 24 to March 3, they worked with members of 1st Canadian Rangers Patrol Group, learning how to move by snowmobile, dog sled and snowshoe in our extreme winter climate which, this year at least, hasn't been that extreme after all.

They also worked on the construction of shelters, lighting fires, living off the land and winter driving as well as conducting trials of their cold weather clothing and equipment.

Part of the exercise was using each others' gear. The Rangers had a fine time trying out the BATUS group's L85 A2 (or SA80) Rifle, General Purpose Machine Gun (GPMG) and the Pump Action Shotgun, while the Brits tried out the Rangers' standard Lee Enfield 303 Rifle.

Arctic Roller isn't exactly a "need" as far as the BATUS group is concerned. Hargreaves calls it a "nice to do.

"It's an all-round super experience."

For the visitors, that even included ice fishing, trapping a rabbit and skating on the ice road to Tuk.

Once in Tuktoyaktuk, they joined up with the Rangers there. They built igloos and watched as the locals showed them trapping skills and ice fishing. Members of the BATUS group were particularly impressed by the way the N.W.T. Rangers lived off the land, taking very few supplies into the field with them.

"When we go on an exercise," said Sgt. Ben Wedgwood, "we take everything with us to survive, supplies, kit and everything. There, they've got one sled, some wood, a tent and a stove and off they go hunting or fishing."

While that may limit the mobility of the patrol in some ways, it makes them very self-sufficient.

Just lighting a fire, even around the Dawson area, can be a challenge, as the BATUS group discovered.

"They all passed," Ranger Sgt. John Mitchell chimed in, adding that Hargreaves was one of the "two brave souls that slept in the improvised shelters" while everyone else enjoyed the shelter of the wall tents.

Back in Dawson, the group had a day's respite before driving south again.

On March 11, they headed for Whitehorse, a little discouraged to be driving in 20 cm of snow, but happy that it had vanished by the time they got to Stewart Crossing.

– 107 –

The Canadian Rangers: More than a rifle and a red sweatshirt

Whitney Lackenbauer

The Dispatch: Quarterly Review of the Canadian Defence & Foreign Affairs Institute, Spring 2010

The Arctic has heated up, as much in political and military rhetoric as in temperature. Journalists warn that a "new Cold War" is brewing over the expectation of new transportation routes and easier access to Arctic resources, citing academic commentators who are eager to highlight the geopolitical complexity of the region and the possibility of future conflict. The short-term horizon, however, is certainly less bellicose than proponents of an "Arctic race" scenario intimate. There is no "use it or lose it" situation for Canada that demands an urgent military response. The Arctic islands are unquestionably Canadian, the Northwest Passage is Canadian waters (although the Americans and others assert that there is an international strait running through it), we are managing boundary disputes with our closest neighbours, and the Arctic littoral states (Canada included) are preparing their extended continental shelf claims. As much as things have changed, they have also stayed the same. As my forthcoming book with Peter Kikkert on *The Canadian Forces and Arctic Sovereignty: Debating Roles, Interests, and Requirements, 1968-1974* reveals, a larger Canadian Forces (CF) presence will not bolster our legal sovereignty position. We have all the sovereignty that we need, and on the front lines of the changing Arctic stand the permanent residents of the region. They have a primary role to play in its defence – a role that is not new, despite the vicissitudes of southern Canadian interest in the region's security.

This past summer, the Chief of the Defence Staff, General Walter Natynczyk, applauded the Canadian Rangers for their "tremendous skills." He was right on the mark. The CF boasts one of the most impressive assets in the North. The Rangers simultaneously serve their country and their communities. They bring indigenous skills and local knowledge about the lands and waters, and serve as an essential bridge between northern peoples, the military, and the federal government more generally. When something is not broken, it is best not to break it. Nonetheless, promised investments to enhance their capabilities and training are well directed as long as they respect the Rangers' longstanding roles and mission.

From 1947 to the late 1960s, the Canadian Rangers did not receive much, if any, formal training. For the last four decades they have been trained by Ranger instructors – fulltime Regular or Reserve Force sergeants or warrant officers who annually travel to communities to work with them and to act as liaisons with the CF establishment. The Rangers are not prepared for combat, but rather to act as the military's eyes, ears and voice in remote regions. By all accounts, they perform these roles admirably. The raucous applause for the Rangers at the closing dinner of Op Nanook, held in Iqaluit last August, shows the high regard with which they are held by southern troops.

The Canadian Rangers are lightly equipped and self-sufficient for operations near their communities. In the early years, Rangers received only a rifle, a hundred rounds of ammunition, and an armband. They still receive the same .303 rifle as they did in the late 1940s, although a long discussed replacement seems to be in the works. One hopes that it will be equally robust and reliable. Since the 1980s, the Rangers are best known not for their armbands but for their venerable red sweatshirts and ball caps. They now receive combat pants, boots, and are expected to receive additional clothing to do their jobs. Ranger patrols are issued tents, GPS units, stoves, radios, and other equipment. The federal government has promised more equipment to support them, including boats to facilitate a marine role. These are all welcome investments.

So too are investments in the future. There are currently 37 Junior Canadian Ranger patrols in the territorial north (1st Canadian Ranger Patrol Group), with more than one thousand young Northerners participating. This program is hugely popular amongst youth, builds self-esteem, and facilitates the sharing of traditional and local knowledge. This encourages sharing of knowledge and skills across generations, and is a clear case of mutual benefit: communities appreciate opportunities to teach young people on the land; and the military is helping to ensure that "skill fade" will not reduce the local knowledge required to conduct effective operations.

In short, if anyone wants to make the case that the federal government has "bypassed the Inuit,"* he should not base it on the Canadian Rangers. They are a clear example of a relationship that the government has actually got right. The Rangers are proof that modest, grassroots measures can assert a national presence over Arctic lands and waters. Because Ranger activities allow Northerners to exercise jurisdiction and control, they demonstrate that traditional indigenous activities continue unbroken to the present and are supported by Ottawa. This is a key pillar of Canada's sovereignty position.

The Conservative government has committed to expand and enhance the Rangers, and the key will be seeing it through. Active recruitment is underway and the Rangers are a key component of the Land Forces' Northern rapid response plans. They will provide local expertise and support for a small land manoeuvre element, which is a logical short-term solution so that Canada can respond to the most probable safety and security issues that Canada is likely to encounter in its Arctic in the next decade. For all the hype about the Northwest Passage, sovereignty loss, and circumpolar conflict, the 2009 Arctic Marine

* Contrast this with Colin Alexander's 24 August 2009 *Edmonton Journal* article "Real Canada has bypassed the Inuit," which incorrectly suggests that Lackenbauer disagrees with Natynczyk's appraisal and considers the Rangers little more than a rifle and a sweatshirt.

Shipping Assessment does not anticipate that it will become a viable trans-Arctic route through 2020. In the meantime, while Canada clarifies its sovereign claims to the extended continental shelf, aligns its Arctic regulatory system for shipping with international standards, and builds air and naval platforms to meet potential threats, the Rangers will continue to play an important role in gathering intelligence, teaching southern personnel, and demonstrating Canada's sovereignty over the land and waters through their daily activities.

With all the alarmist rhetoric about Arctic security and sovereignty swirling around these days, the real challenge is to manage our expectations of the Canadian Rangers as a vital northern asset. Residents of remote Northern communities already serve in the Rangers at rates more than five times higher than that of southern Canadians serving in the Regular Forces or Primary Reserves. The Rangers are a flexible, inexpensive and culturally inclusive means of "showing the flag" and provide tangible support to military operations and training. They deserve both our gratitude and respect, as General Natynczyk suggests. The Canadian Forces must be careful not to set the Rangers up to fail by asking too much of them or trying to over-militarize them to face a theoretical enemy that is unlikely to challenge our Arctic sovereignty and security in the near future. Northerners will not stand for it, despite their continued willingness to stand on guard for thee.

– 108 –

Rangers Out in the Field

Terrace Standard, 31 August 2010

Northwest Canadian Rangers not only received some valuable training over the weekend, it was also tied to a historical event involving the crash of an American nuclear bomber in 1950.

At Gil Island on the coast, where the B-36 bomber's crew bailed out because of engine failure during a training mission, and at Kologet Mountain, 178 km north of here, where the then-pilotless bomber eventually crashed, rangers practised search and rescue techniques, radio communications and leadership.

The weekend was called Exercise Skookumchuk and the exercise locations were chosen because of their historical significance, says Warrant Officer Dan Hryhoryshen, a Victoria-based training officer and who was in charge of Skookumchuk.

Hryhoryshen was in charge of organizing the exercise and says it was an opportunity for rangers to put skills they've already learned into practice.

Rangers converted the Terrace Rod and Gun Club into an operations base camp before setting off to their assigned locations, either on Gil or Kologet.

Three rangers were placed on Gil Island in advance to be parachute "casualties" that other rangers had to then rescue following proper procedures.

The rangers in charge of the search and rescue were boated out and dropped off on the northern tip of Gil Island and tasked with finding the placed victims only using a grid as a reference to navigate the co-ordinates. And once finding the victims the rangers had to find a way to bring them back to the beach for pick-up.

With four civilian helicopters chartered for the exercise, it was one of the largest recent ventures.

"An exercise of this magnitude doesn't happen that often, this is a bit of a big show," said Hryhoryshen.

Rangers who were dropped off on Mount Kologet had specific tasks to complete including securing the site, searching, finding and locating as well as practicing communications back to base camp using high frequency radios with help from the Canadian military from the Nanaimo base.

This meant that the rangers had constant contact 24/7 over the weekend with the base camp at the rod and gun club.

"Other elements they will practice will be...leadership," said Hryhoryshen. "Also to practice command and control, the ability to think on your feet, and to get the job done without supervision."

He also added that two rangers were chosen the day before the exercise, without notice, to be in charge. This was followed by a briefing of the tasks at hand.

This isn't the first time Mount Kologet has been used by the Canadian Rangers. They even played a roll in cleaning up the crash site and making sure there were no hazardous materials scattered around.

"[Since then] the Canadian Rangers developed exercises to practice skills, a number of years ago they went up there and did a survey to make sure it was clean and safe," said Hryhoryshen. "We went back five years later and did an exercise...it's an exercise in getting there. It's very interesting to look at, there are bits and pieces (of the plane) that look brand new, could be made yesterday, other stuff looks 50 to 60 years old."

The plane did have a nuclear weapon, but not a plutonium core. It was jettisoned over the Pacific before the crew bailed out.

Fortunately before the plane crashed back in 1950 the bomb was dropped somewhere around the Hecate Strait. Only the bomb's highly explosive material detonated, as the plutonium core was not installed in the bomb, no nuclear explosion took place.

There were seven rangers who participated from Terrace with other northwest rangers coming from Sandspit, Port Clements, Bella Coola, Kitimat and one from Gold River on Vancouver Island.

Hryhoryshen says that Terrace has a very active group of rangers with 427 registered in all of B.C.

– 109 –

Embedded with the Canadian Rangers

Tim Querengesser

Up Here, vol. 26, no. 7 (October-November 2010)

If Canada's Arctic sovereignty has a brand, it's the red Rangers hoodie. But what does this citizen militia do to ensure the Northern frontier stays ours?

The wave slaps the boat into a high-speed salsa dance. "Slow down!" screams Ranger Corporal Lori-Ann Lennie, beside her niece, Kayla MacCauley, who's beside me. That's me, behind the jet boat's steering wheel. Now, everything starts to happen extremely fast. As the boat gyrates like Elvis and threatens to chuck us into the Mackenzie River, I reach to my left and accidentally push the throttle fully open. The engine roars like God impersonating a grizzly bear and sends the nose skyward. Now, properly scared shitless, I smack the throttle shut, grip the wheel and mutter a prayer. The boat's nose crashes down with a "Thump!" but, thankfully, stays above the waves.

With the engine cut we begin to drift silently with the current. And having nearly killed the strangers sitting beside me, I wait for a scolding. "Oh my god!" Lori-Ann says, giggling. "Do you even know what yer doing?!"

No. No I don't.

I didn't expect my first experience with the Canadian Rangers to be driving a jet boat. In fact, I didn't expect to drive a boat, period, even if Lori-Ann's boyfriend, David, said he wanted a man piloting his boat and even if I happen to fit that description. But 20 minutes into my four-day embed with the Rangers and it was already clear that my expectations were way out of touch. I'd expected searches for Russian bombers, or foreign submarines, or some macho assertions of sovereignty, Hoo-Wah! Now, those all seem hopelessly naïve.

As I try to park the jet boat against the rocky riverbank, I manage to point it backwards, then sideways, struggling to negotiate the Mackenzie's strong current. My mind flashes back to my five-minute test-drive – performed to calm David's legitimate concern I have no idea what I'm doing in a jet boat. "Landing is the hardest part," he warned. It's proving true. The other Rangers giggle and Lori-Ann is electrified. "You see," she yells. "I could have driven the boat."

Yep.

It's likely a majority of Canadians share my misconceptions about the Canadian Rangers. Most probably have a vague notion that this 4,400-strong citizen militia, 85 percent of which is aboriginal (and predominately Inuit), does something to protect Canada's Arctic sovereignty, but couldn't tell you how. To answer that question, I've come to Tulita, a Dene community of 550 located near the Arctic Circle, and home to one of 165 Canadian Ranger patrol groups.

The squad is headed into the wilds for their yearly patrol. Now, they'll be the first to tell you they're not on the front-lines like some of the more northerly Ranger patrols. Yet four days spent doing things such as nearly dying in a boat, relaxing, joking, and hunting animals, not Russians, gives me time to ponder. How do the Canadian Rangers help protect our sovereignty?

Wilfred Lennie, a former bureaucrat, sits to rest on a plastic cooler. His tininess has earned him the nickname "Peanut," and at 62, he's combined this disarming nature with a sharp wit to become the group's comedian. Between jokes about the "bushman" he says is stalking us, he wisecracks that Canada should send him to Afghanistan "to go get bin Laden." When I ask him about joining Rangers he stops kidding. "Travelling on the land, doing these patrols, that's what I like," he says.

All around Wilfred, 14 other Tulitia Rangers are busy erecting a tent city. There is Lori-Ann, a 45-year-old office manager by day, who stands alongside the six other women on the patrol, giggling. There is William and Joe Horassi, 60 and 67, who speak in North Slavey as they collect telephone-pole sized driftwood from the riverbank or cut branches from the surrounding fir trees to build tent frames. There is Ranger Sergeant Eddie McPherson, the group's elected leader, who inhales a pack of cigarettes seemingly every three hours. And there are Clayton and Ranger Corporal Paul MacCauley, 30 and 25, two intense brothers with wide Jack-O'-Lantern smiles.

"Afterwards you want to go out?" Clayton asks Paul, as he cooks some frozen moose they've brought along.

"Yeah, there's fresh tracks down by that island," Paul says.

"I just want to go and shoot a moose!" Clayton says, wide-eyed.

The camp is set on a river bend overlooking a spot, everyone says, "is good for moose." This isn't a very military-like camp and the patrol's itinerary reflects that: Day one, arrive. Day two, target practice. Day three, hunting. Day four, go home. Add "joke and visit" and you have an accurate picture.

The Tulitia Rangers have packed for comfort rather than mobility for their four day patrol. From their boats they unload oil-drum wood stoves and chimney pipes, coffee thermoses, cribbage boards, camp chairs, gas lamps, a plastic tub with a hind-quarter of moose, boxes filled with industrial-sized stores of macaroni, sugar, coffee-whitener, Kool-Aid, canned meat and lard, and the biggest surprise, black garbage bags containing foamie mattresses – complete with bed sheets.

What army will eat all this food, I joke with Lori-Ann. "I don't know," she says, "but I know who's going to cook it."

With the six-tent camp finally complete, Eddie gathers everyone up. "There's nothing else to do for the night, so if you go in a boat, be back by 9," he says.

"When's supper and who's going to cook breakfast?" Paul shoots back, from within the crowd. "You're supposed to take care of us Sergeant. You're the big daddy of the bunch." Everyone giggles.

I seek out Sergeant Francois Tremblay, the only regular Canadian soldier on the patrol. Francois, from Quebec, did a tour in Afghanistan before joining the Rangers, and he's here as an instructor. He travels throughout the territories helping to oversee 56 Rangers patrols in the first Canadian Ranger Patrol Group, one of five in total across Canada.What now? I ask him. "Now?" he says, in his overly-honest broken English. "Relax."

During the Second World War, Ottawa realized it couldn't defend its own frontiers from threats from the Japanese in the northeast, and later, from Russians in the Arctic. Silently, it opened the Northern door to the Americans. The Alaska Highway, countless Yukon airfields, air and weather stations at Iqaluit, Eureka, Alert and Resolute, and the Canol Pipeline were all initially paid for and built by Americans.

After the war, as its inabilities became more widely known and embarassing, Canada got more serious about keeping watch over its Northern frontier. As a grand gesture it bought the Canol Pipeline back from the Americans, even though the pipe didn't really work. And it followed the United States' lead in creating a Northern military force composed of volunteers.

Since 1942, when the Japanese bombed Alaska's Dutch Harbour and occupied several Aleutian Islands, the Alaska Territorial Guard – a volunteer militia composed almost entirely of native Alaskans – had been keeping watch over the soon-to-be state. And in Greenland, the Danes had created Slaedepatruljen Sirius, an Inuit sledge patrol tasked with spotting unwelcome German weather stations. So in 1947, Canada created the Rangers, from the ashes of the Pacific Coast Militia Rangers, a volunteer force that had been created to scout the west coast and Yukon during the war for possible incursions by the Japanese.

What's happened since is telling. The Alaska and Greenland patrols have been absorbed by the regular military. The Rangers, on the other hand, have carried on much the same. They're still issued .303-calibre, Lee-Enfield bolt-action rifles, designed in 1895. They still get comparatively little training or equipment (to get to this patrol, the Tulita Rangers used their own boats, which Canada rented from them). They're still volunteers – still paid only when on training or patrols, and still given just 200 rounds of ammunition a year.

A Ranger will never shoot at anyone while on patrol. Instead, they use their rifle to hunt as they explore the Northern wilds. If they see, say, a foreign submarine, aircraft, wildlife poacher or a chemical spill, they are trained to document it and inform their headquarters in the south. If someone goes missing on the land, they're called upon to help find them.

Today, despite the throwback feel of the group, the Rangers are expanding. By 2012, it's estimated they could top 5,000 members.

For anyone still sleeping, morning on day two comes at the crack of Master-Corporal Benny Doctor's .303. Doctor fires on a flock of geese flying by the camp, with Sergeant Francois Tremblay beside him, smiling broadly. Eddie's there, too, with his seemingly five-gallon "Bud Light" coffee mug already full. "Get the machine gun ready for the next ones," he says.

Today is target practice day – on a range created on a flat, dry riverbed, firing at targets held up, like the wall tents, with improvised frames tied together with twine. The targets, used throughout the Canadian military, evoke a long-forgotten enemy: A German soldier rushing at you.

"What day is it?" someone asks, as the range is set up."Sunday," Eddie replies, grinning. "We have to interrupt this shooting range to have mass." Behind the firing line people are sitting on camp chairs and ammunition boxes. Here, Wilfred is holding court and Lori-Ann and several others are in hysterics. "They should send us all to Afghanistan, for bin Laden," Wilfred says."We should put him in white clothes and everything," Lori-Ann says, giggling. "He'll look just like bin Laden."

At the firing line, though, things aren't as enjoyable. As eight shooters dial in their sights, including Joseph, who smokes a cigarette while firing, Stephanie MacCauley starts complaining. Her rifle isn't accurate. "I gotta shoot left handed I guess. It's not shooting nothing."

Over the course of the day, Paul and Francois work on Stephanie's rifle, at one point throwing a stick in the water, firing at it and watching the splash to determine how to adjust it. Afterwards, Paul declares: "That gun's fucked."

I talk with Lori-Ann about the links between the Rangers and the military. To her there aren't many. In 2007, she was part of Operation Nanook, where the regular army came to nearby Norman Wells and the Rangers were employed as wildlife patrol. One night, at the Legion – after she'd hustled them all for their money at the pool table – things got a bit serious. "It was a lot of fun until all those army guys started fighting in the Legion," she says. "They just ruined it." Instead, Lori-Ann sees benefits for herself and her family. Her daughter has recently joined the Junior Rangers.

Eddie, however, sees definite links. A Métis construction worker, he joined the Rangers nine years ago with clear reasons. "I seen, you know, these guys trying to do something – you know, sovereignty, eyes of the North, all this stuff," he says. "I wanted to be part of it. The other day, when that Russian plane was coming over – who better to protect the North than Northerners?"

But, he concedes a few frustrations, too. "The worst part of all this is ... we have these antique guns," Eddie says. "It's hard to get parts for them. A lot of them have bent barrels." Challenges also include wall tents sent north without holes for wood-stove pipes and a general shortage of supplies. The two tents the Tulita group has are expected to house more than 30 members. The only solution is fundraising. "Although we get it from [Canada] we don't get enough, so we have all our own [gear]," Eddie says. "Our next goal, if we ever make money at bingo again, is to buy some more tents."

Kayla is a quiet presence at the range. After firing, I talk to her about her reasons for being in the Rangers, which she joined at 19.

The Rangers were "just something to do, because there's nothing like really much to do in Tulita," Kayla says. The more we talk, the more it's clear this group is a way for Kayla to keep in touch with her culture, too. She tells me her family used to go out on the land a lot. "As I got a bit older, you know, too much drinking in the family and stuff, and I guess we just never went as much," she says. "This is probably the only time that I go in the bush, is with the Rangers, and the [Tulita] spring hunt."

Surprisingly, Wilfred – who I chat with after the shooting is done and we're back at camp – has a similar story. A lot of kids in Tulita drink, and some kill themselves, he says. Drinking is what he did, too, before he retired from a 25-year career with the territorial government in 1997. That same year, when the Tulita division of the Rangers was created, Wilfred signed up and got sober. "When you're out here, see how beautiful it is, sometimes you think, 'I want to live longer,'" he says. "But the kids aren't coming out on the land."

On day three, Clayton, Paul and Joseph are finally let loose on the river. The night before they tasted first blood, shooting two ducks that swam past the camp and de-feathering them in a flurry beside the river. Today, like everyone else on the patrol, they want the big prize: a moose.

We head downriver in a Lund fishing boat, with Clayton steering, Paul near him, and Joseph sitting at the bow. We take a right down a small creek where, at the top of a near-vertical riverbank, a hunter's cabin overlooks what the guys call Harry's Island. We climb up and have a look into the willows. "I just wanted to get away from that camp," Clayton says, lighting a cigarette and relaxing in a chair.

I ask why, if he doesn't like the camp, he joined the group. "I thought the Rangers were that soldier-boy stuff," he explains. When he found out they weren't, he signed up. "I did it to go hunting. It's good to get out with a bunch of people and go hunting."

Clayton and Paul both have full-time jobs in Tulita, with the local government. They've come home: The two used to live at a mining camp north of Yellowknife, partying in the city on their off time but feeling disconnected. "In the spring-time in Yellowknife I saw a flock of geese," Paul says, scanning the horizon for moose. "I just said, 'Man, I want to come home.'" Both brothers returned that year for a relative's funeral and never left. Now, Paul says he barely drinks at all.

We wait for a few moments, breathing in the beauty of the Mackenzie River valley, the trees exploding into the florescent yellows of autumn, the water splashing in the creeks, the mountains looking like purple-carpeted blobs in the distance.

Clayton grows tired – we haven't seen a moose all day. "C'mon, let's keep looking," Paul says. "Do you want to go back to that camp and sit in tents? How boring."

As we continue hunting, it strikes me that there are two vastly different ways we can assert sovereignty in the North. One is to militarize the place, to prepare a highly-unlikely future battle over the Arctic. The other is to link the country's broader goals with those of the local people. They, like Ottawa, want to protect the North. Now, if we could only give them some more tents and better rifles in the cause.

– 110 –

General Visits 'The Eyes and Ears of Canada'

Jason Unrau

Whitehorse Star, 12 January 2011

The Canadian Rangers, a more-than 4,000-strong militia composed of volunteer residents from the country's northern communities, are expanding, but there are no plans to upgrade the Rangers' standard issue, vintage Lee-Enfield rifles.

Canada's top soldier, Chief of Defence Staff General Walt Natynczyk, said as much during a brief stop in the Yukon on Tuesday to visit with Rangers from Haines Junction, Whitehorse, Carcross and Carmacks.

"Over the past five years, this is an issue that's come in and gone out so many times, because we have folks, mostly from the South, who want to give the Rangers a newer, more modern weapon," Natynczyk said.

"But the feedback we get from many Rangers, depending on who you talk to, they want a simple weapon.

"And the Lee Enfield .303 rifle that the Rangers have, although it's old, it's one of the most reliable, simple and accurate weapons, that's ever been designed."

Officially adopted as the British Army's standard issue rifle in 1895, the Lee-Enfield is still used by some Commonwealth nations' armies, but for a modern military like the Canadian Armed Forces, the weapon has long been retired.

While some Rangers have complained about the difficulty in finding replacement parts for the Lee-Enfield .303, Natynczyk said he has heard the opposite from other Rangers who appreciate the gun.

"I still remember Sgt. (Allen) Pogotak from Resolute Bay at Rideau Hall, when he was being given an award by the Governor General," Natynczyk recalled.

"He said to me, 'Sir, you know, you can take this weapon, it can be dropped in the ocean, you pick it up and shoot and it fires and fires true. And when anyone in my patrol breaks this weapon, I can go on the Internet and order the parts, and it's delivered in a week.'"

And expanding the Rangers' scope of service beyond regular patrols and joint exercises with the Canadian Armed Forces - a 2009 Senate report on Arctic sovereignty suggested that Rangers be turned into combat-ready units - is not being pursued.

"The government's intent to grow the Rangers' numbers, that was part of the Canada First Defence Strategy," Natynczyk said of increasing membership, rather than increasing responsibilities.

"But it's a question of how the communities want to expand. We can't just go out there and say, 'Hey, we want everyone to join the Rangers.'"

Members of Joint Task Force North, based in Yellowknife, N.W.T., also accompanied Natynczyk on his brief two-day tour of the hinterland.

"The Rangers are the eyes and ears of Canada in all of these communities," the general said of the volunteer force established in 1947 with the motto *Vigilans* or The Watchers.

"And everytime I'm out here, I'm just humbled by their skill and their knowledge."

There are approximately 4,200 Canadian Rangers in 163 patrols across Canada, and 85 per cent of Rangers are aboriginal.

– 111 –

New Canadian Ranger Patrol a Source of Pride for Lac Seul

Sgt Peter Moon

3 CRPG *Newsletter*, May 2011

A graduation parade marking the opening of Northern Ontario's newest Canadian Ranger patrol was "a very proud day for us," said Chief Clifford Bull of Lac Seul First Nation.

"I think it's awesome that we now have this patrol here in our community," he said. "It's something that we've always wanted and feel we have needed. It is a tremendous day for us.

"These Canadian Rangers will be setting an example and acting as role models for our young people, who will be looking up to them. It's exciting to hear that we can look forward to getting the Junior Canadian Ranger program established in our community, too."

The opening of the Lac Seul patrol in February brought the number of communities with Ranger patrols in Northern Ontario to 19 and the total number of Canadian Rangers in the province to 500.

"The graduation parade was great," said Major Guy Ingram, commanding officer of 3rd Canadian Ranger Patrol Group.

"Everyone seemed to have a family member here or a friend and there was great support from the community. It was good to see them taking pictures and getting involved."

Brigadier-General Fred Lewis, the reviewing officer and commander of Land Force Central Area (the military name for the army in Ontario), told the graduates they would be a valuable addition to the Canadian Forces' resources in Ontario's Far North. He presented Sergeant Brad Ross, the new patrol leader, with his sergeant's stripes.

"It feels great being the new patrol sergeant," said Sergeant Ross. "I was very honoured and excited at the same time."

He is a Lac Seul band councillor and auxiliary police officer. He was formerly a Canadian Ranger with the nearby Muskrat Dam patrol.

– 112 –

Rangers Train 700 Troops in Winter Survival

Sgt Peter Moon

The Maple Leaf, 11 May 2011

Canadian Rangers from four isolated First Nations in Northern Ontario provided winter survival training for more than 700 troops over a two-month period this year.

"That's a remarkable achievement," said Master Warrant Officer Robert Patterson, 3rd Canadian Ranger Patrol Group's Canadian Ranger sergeant major. "Every time the troops train with the Rangers, they are just in awe of the whole Ranger mystique and background. They are very enthusiastic about working with them and they pay close attention to what the Rangers are trying to pass on to them."

The Ranger instructors were from Constance Lake, Fort Albany, Kashechewan and Moose Factory. They taught traditional winter survival skills during three major winter warfare exercises for members of The Royal Canadian Regiment, and 32 and 33 Canadian Brigade Groups (CBGs). Participants included Regular and Reserve Force personnel, cadets from Royal Military College of Canada, members of the New York National Guard, and an officer of the Australian Army.

Training activities included ice-water rescue, building improvised shelters, fire-starting (including emergency signal fires), snaring, wild-food preparation, axe- and knife-sharpening, safe use of chain saws, snowmobile maintenance, and other useful survival skills.

Private First Class Joey Delancey of the New York National Guard was in awe of the Rangers after he snared his first rabbit during training with 33 CBG in the bush near Gogama, Ont. "I'm from New York City," he said. "I've only ever seen a rabbit inside a pet store. I've never seen a wild rabbit before. I think the Rangers are great. They have a lot of knowledge."

Ranger Amos Ferris, an Oji-Cree, was one of 17 Rangers taking part in the Gogama exercise. He said he was proud to be able to pass on some of the knowledge he learned from his grandfather. He snares to put food on his family's table and was surprised by the resistance of some of the female soldiers to learning how to snare.

"Most of the ladies said they were against animal cruelty, that they didn't want to eat an animal, but I don't consider it animal cruelty at all," he said. "It's a way of survival and it's necessary if you're ever caught in the bush and need to eat to survive. I told these ladies they should learn, because they might need these skills one day."

"The value the Rangers bring to these exercises," MWO Patterson said, "is their traditional knowledge in how to survive in the bush in the extreme cold of the Canadian winter. By and large, that ability has been lost in the Canadian Forces over the last decade or so and we are regaining it with the help of the Rangers.

"Over the past five years I've seen the Rangers evolve into very good instructors. They always had the skill set. It was just a matter of them being able to transmit that knowledge. They've got the confidence to do it now. They are in great demand and they do a great job."

– 113 –

Rangers Assist during Forest Fire Threat to First Nation

Sgt Peter Moon

The Maple Leaf, 20 July 2011

Six Canadian Rangers played an important role in the evacuation of elders, pregnant women, children and the sick when a major forest fire threatened an Ojibway community in northwestern Ontario.

"The smoke from the fire was bad; it made it hard to breathe and it got in your eyes," said Master Corporal Mervin Masakeyash. "We were all over the place, telling people about the voluntary evacuation, counting people who were leaving—and those who were not—and getting them on planes."

About 530 of the 900 people in Mishkeegogamang First Nation took advantage of the voluntary evacuation and left for temporary accommodations in Sioux Lookout, Ignace, Geraldton, Pickle Lake and Thunder Bay until the fire was brought under control in early July.

The Rangers "did everything from giving assistance to the chief and council in the command post, answering telephones, spelling people off for their breaks, and stuff like that," said Warrant Officer Mark Kendall, a 3rd Canadian Ranger Patrol Group instructor who was in the community when the crisis occurred. He was joined by 3 CRPG instructor Sergeant Barry Borton, who flew to Mishkeegogamang to assist.

"The Rangers went door to door—sometimes alone, sometimes with a police officer—to tell people about the evacuation arrangements and pass out the voluntary evacuation notices," WO Kendall said. "They told people of the dangers of the smoke and that the priority was to get children, pregnant mothers, the elderly and people with respiratory problems out first. They showed the people of Mishkeegogamang the value of having a Ranger patrol in their community."

MCpl Masakeyash and the Rangers accompanied 60 of the evacuees when they flew to the northwestern Ontario emergency reception centre in Geraldton, where they were put up in the town's arena.

"We were all over the place, telling people about the voluntary evacuation, counting people who were leaving—and those who were not—and getting them on planes," MCpl Mervin Masakeyash explained.

"We kept people occupied," he said. "I organized events like baseball and floor hockey and swimming. We took turns being on duty at night. Everything went very well."

Rangers have played lead roles in several emergency evacuations in Northern Ontario during forest fires and spring floods in recent years. When tainted water forced the evacuation of Kashechewan in 2005, Rangers from several Northern Ontario communities assisted.

"What the Rangers did in Mishkeegogamang goes to show the level of dedication that the Rangers have to their community and to the Army in an emergency," said Captain Mark Rittwage, officer commanding the Canadian Rangers in Northern Ontario. "Just as in Kashechewan, there were Rangers who volunteered to support their community."

– 114 –

The Canadian Rangers: A CF Touchstone in Canada's Remote Regions

Sentinel Staff, with information from Joint Task Force North and the Canadian Ranger National Authority

The Sentinel (Joint Task Force North), July 2011

They come from all walks of life. Some are mayors, chiefs, teachers or businessmen, and some earn their keep off the land by hunting and fishing. They are of all ages and both genders; most are First Nations people - but many are not. But they all share some invaluable traits. An extraordinary knowledge and sense of the vast and rugged spaces of our land, primarily in the North but not exclusively so, as they are also found almost everywhere Canada has a coast line, from the West Coast, to the shores of Hudson Bay to the St. Lawrence Estuary to the shores of Newfoundland and Labrador.

They operate in conditions that no reality show could emulate, where the weather is often cold, cruel and unpredictable, and where the land is often covered with snow and cloaked in darkness, and inhabited places are few and far between.

These are our Canadian Rangers whose over 4,000 members, scattered in patrol groups at over 170 locations throughout Canada, provide a strategic resource to the Canadian Forces (CF) out of all proportion to the investment in them, and who work on behalf of Canada and the CF but above all for love of country.

The Canadian Rangers were founded on May 23, 1947 to provide a permanent presence throughout the North and to assert Canadian sovereignty through surveillance, most often by means of sovereignty patrols. With 40 percent of our landmass in the territories, 162,000 kilometres of Arctic coastline and 25 percent of the global Arctic, Canada is undeniably an Arctic nation. Sovereignty across this vast expanse is not limited to air patrols or missions by Canadian Coast Guard and Navy ships, or periodic exercises by the Canadian Army, but also by Canadians patrolling on the land year round - often on snowmobiles, whose knowledge of the land can quickly discern if something is amiss.

Identified by their signature red sweatshirt and red baseball cap with a Ranger crest, Rangers help to advance the four pillars of Canada's Northern Strategy by enhancing sovereignty, protecting the environment, encouraging economic and social development and building confidence for the residents of the North to take on provincial-type responsibilities for land and resource management.

They achieve this by receiving important CF training, support and responsibility, and by demonstrating sovereignty through their presence and support in the North. They also work with other CF members across Canada, showing them the skills and ways of life that are unique to Northern environments and resources. This collaboration and knowledge-sharing by the Canadian Rangers enables the CF to ensure the safety, security and defence, as well as the environmental, social, economic and cultural protection of this isolated, exceptional landmass and its communities.

At a recent awards ceremony presided over by the Chief of the Defence Staff, General Walter Natyncyzk, the Honourable Eva Aariak, Premier of Nunavut, delivered these praising words to the Canadian Rangers: "Each of you serves as the military's eyes and ears in the North, helping the Canadian Forces to defend our country and protect our national security. It is your knowledge of ways of the land, hunting, weather conditions and your endurance and strong surveillance instincts that make it all possible."

Operating as the eyes and ears of the Canadian Forces, Canadian Rangers can be counted on to provide military support in times of community crisis, to act as the advanced guard in the remote areas of the North, and to participate in several capability-sharpening operations and exercises conducted by Canada Command every year in the North.

"There are large parts of this country where we don't have a permanent Regular Force or Primary Reserve presence," said Dr. Whitney Lackenbauer, associate professor and chair of the department of history at St. Jerome's University in the University of Waterloo. "The Canadian Rangers ensure that the Canadian Forces are well represented all across this country in all of its geographical and cultural diversity."

This is a mutually supportive relationship where the presence of the Canadian Rangers greatly helps the people of these communities and it also helps Canada Command to meet its objective to "Protect and Defend" all of Canada.

These important members of the Canadian Forces are not employed in combat or overseas operations. However, they have played key roles in the evacuation of several Aboriginal communities threatened by forest fires, tainted drinking water, spring flooding, avalanches and other domestic emergencies. The honed search and rescue skills of these Canadian Rangers have made the difference between life and death for the many people who go missing in this vast landscape.

Canadian Rangers take pride in serving their people as well, and their skills are supported and encouraged by the CF. "They're serving their country at the same time as they're serving their community," said Dr. Lackenbauer.

Canadian Rangers are provided with patrol-specific training that includes first aid, in-service weapons use, guiding or scouting, navigation using maps, compasses and global positioning systems, traditional and modern survival skills, operations in support of Regular Force units, basic military drills, surveillance, search and rescue, North warning system patrolling and patrol leadership, and administration.

Canadian Rangers are organized into Patrol Groups throughout the Land Force (Army) areas across Canadian Provinces and in Canada Command's Joint Task Force North. The patrol groups tend to be provincially oriented, apart from 1 Canadian Ranger Patrol Group, which covers the whole of northern Canada above the 60th parallel. The patrols tend to be centred on remote communities and are frequently named after the town or village they are from.

Located mostly on the coast and in the north, the Canadian Rangers form 173 Ranger patrols in Canada, divided into five geographical areas defined as Canadian Ranger Patrol Groups (CRPGs). Each CPRG has a headquarters and a staff that is responsible for overseeing the Canadian Rangers and Junior Canadian Rangers in their respective area as follows:

1 CRPG is headquartered in Yellowknife, NWT, reports to Joint Task Force North (JTFN) and operates throughout Nunavut, Yukon Territory and the Northwest Territories;

2 CRPG reports to Land Forces Quebec Area (LFQA) and covers the entire province of Quebec;

3 CRPG reports to Land Forces Central Area (LFCA) and is responsible for Ontario;

4 CRPG reports to Land Forces Western Area (LFWA) and covers British Columbia, Alberta, Saskatchewan and Manitoba; and,

5 CRPG reports to Land Forces Atlantic Area (LFAA) and operates in Newfoundland and Labrador.

The Government of Canada has increased its emphasis on Northern sovereignty, placing a greater emphasis on the contributions of the Canadian Rangers. This recognition of the tremendous contributions the Canadian Rangers has resulted in a commitment to increase the size of this extraordinary force to 5,000 Canadian Rangers in 178 patrols by the end of 2012.

– 115 –

Visiting the Canadian Rangers

Hon. Peter MacKay

Minister of National Defence's Speech, Whitehorse, 24 July 2011

It's a pleasure to be with you here in Whitehorse, in the heart of the Yukon.

After being in the sweltering heat in Ottawa for several days this past week, it's a wonderful treat to be with all of you in such a picturesque location.

Of course, it really comes as no surprise that Boyle Barracks is situated in a truly sublime setting. It is, after all, a part of Canada's North.

You are all blessed to live in a land of stunning landscapes, stark beauty, and [unparalleled] allure.

It's a land that has long captivated the imagination and the affection of Canadians from across the country, and of people from around the world.

And although the Yukon is a corner of the globe rich in resources and overflowing with astonishing vistas, it's important to be mindful that you – our Canadian Rangers – are charged with protecting the most important and vital part: Yukoners themselves.

You are the *vigilans* - "the watchers"...

...you are self-sufficient and mobile...

...your knowledge of the environment is unmatched...

...your relationships with your neigbours are strong and longstanding...

...and in many parts of the north, *you* are the Government of Canada's most visible presence.

That's a great honour, and also a great responsibility.

As you know, the Government of Canada has made the protection and defence of Canada's North a cornerstone of our agenda. In 2009, we released our Northern Strategy where we outlined priorities of economic and social development, environmental protection, governance, Arctic Science and Technology and of course, exercising our Arctic sovereignty.

The work of the Department of National Defence heavily contributes to this last priority – which means we pay special attention to work being done on behalf of the Government of Canada in places like Whitehorse, Resolute, Iqaluit, Arviat, James Bay and Kuujjuaq.

Bearing that in mind, I'm delighted to be here today to acknowledge the incredible dedication you display in carrying out your work in support of Canadian Forces operations in the North, and in support of your communities.

The assistance you provide to Canadian Forces domestic operations across the country, as well as sovereignty operations in the North, is nothing short of invaluable.

Just ask the five hundred soldiers from Valcartier who, thanks to the support of the Canadian Rangers, were introduced to Arctic survival skills earlier this year in James Bay.

The value of your insight is tremendous, and the importance of passing it along to your colleagues is becoming more and more pronounced. I have every confidence that your knowledge will have a lasting impact on the growing expertise of our Canadian Forces.

Your activities are essential to empowering the rest of the Canadian Forces – as well as other Government departments – to operate safely and effectively in the North.

As I've mentioned, the Rangers have contributed an enormous amount of local knowledge, expertise and mentorship.

The assistance that the Canadian Rangers provide daily is important, for example, in the joint work done by the Department of Fisheries and Oceans, the RCMP and the Canadian Forces through search and rescue operations across Canada. In fact, scores of such operations have benefitted from your support in the last few years alone

Your assistance is also deeply appreciated by federal, provincial or local authorities elsewhere in Canada – sometimes in places you would least expect it.

Some Canadian Rangers, for example, went as far south as London, Ontario, to provide assistance during last December's snowstorm in the region.

Canadian Rangers also helped evacuate 1,200 residents from the threat of forest fires at Wollaston Lake in Saskatchewan last month…

…and most recently Canadian Rangers have helped assist with forest fire evacuations that have ravaged remote communities across Northern Ontario.

At most recent count, Canadian Forces members have airlifted literally thousands of people from the small Ontario towns of Cat Lake, Deer Lake, Fort Hope, Sandy Lake, Mishkeegogamang, and Kingfisher Lake. Many of these were First Nations communities.

Thanks to the relationships that have been forged between members of Ontario's 3rd Canadian Ranger Patrol Group, your colleagues served as trusted envoys for the evacuation of ill and ailing residents of these remote First Nations communities.

Beyond your operational contributions, the Rangers are also a key element of the Government of Canada's ability to exercise our sovereignty in the North.

Your role in conducting and supporting surveillance and presence patrols year-round is an essential part of safeguarding Canadian territory.

And you always provide essential assistance to the major sovereignty operations that are held every year – Operations Nunalivut, Nunakput and Nanook.

In fact, some Rangers are currently preparing themselves to participate in this year's edition of the whole-of-government Operation Nanook, set to last all of August.

Operation Nanook is an excellent way to improve the Canadian Forces' command and control capabilities in the North…

…to enhance interdepartmental and intergovernmental coordination with the Canadian Forces' many partners in the region…

…and even to improve interoperability and understanding with allies such as the United States and Denmark.

That's exactly what we hope to accomplish through this summer's Operation Nanook – the largest edition of the operation in the history of its existence.

Ladies and gentlemen, our Government recognizes the essential nature of all your work, and as such we understand the crucial role that the five Canadian Ranger Patrol Groups play on behalf of our country.

That's why, four short years ago, Prime Minister Harper and I announced that the Canadian Rangers would undergo a major expansion.

As he rightly stated at the time, "the Rangers are a tangible expression of Canada's ability to defend its northern lands."

By growing the Rangers, our aim is to further increase the Canadian Forces' reach and effectiveness in Canada's remote and sparsely inhabited areas, including the North.

In 2007, when this expansion was first announced, there were 4,100 Rangers, and 161 Ranger patrols.

Since then, existing Ranger patrols have been expanded, and new patrols have been added.

I'm pleased to announce that nation-wide, the Canadian Rangers have grown to a current strength of nearly 4,700 Rangers and 173 patrols.

What's more, we are on track to reach our ultimate goal of 5,000 Canadian Rangers and 178 total patrols by the end of fiscal year 2012.

This will include over 1,700 Rangers among 59 patrols in the North.

This means that there will be more Rangers, in more parts of Canada, doing more of the great work that Canadians have come to expect of you.

All of this growth is allowing the Canadian Forces to maintain a stronger presence across Canada.

And it represents the Government's firm commitment to protecting the prosperity – and the people – of Canada's North.

You and your colleagues figure prominently in our plan to exercise Canada's northern sovereignty and ensure the safety and security of northern Canada.

As such, it will give me no greater pleasure than to join with your commanders in a few moments to single out a few of you whose steadfast service and unwavering loyalty has earned special recognition.

Once again, it's a great honour to join with all of you here in Whitehorse – in the Land of the Midnight Sun…

…to salute your accomplishments…

…to applaud your dedication…

…and to celebrate your future.

I thank you once again for all the incredible work you do on behalf of your communities and on behalf of Canada.

Thank you.

– LIST OF ACRONYMS –

2Lt	Second Lieutenant
.30-30	.30-30 lever action rifle
.303	.303 Lee-Enfield bolt action rifle
AC	alternating current
Alta.	Alberta
APC	Armoured Personnel Carrier
A.S.T.	Atlantic Standard Time
BATUS	British Army Training Unit Suffield
BGen	Brigadier-General
Brig. Gen.	Brigadier-General
B.C.	British Columbia
C	Celsius
CADPAT	Canadian Disruptive Pattern
Cal	calibre
Canol	Canadian Oil
Capt/Capt.	Captain
CB	Companion of the Order of Bath
CBE	Commander of the Order of the British Empire
CBG	Canadian Brigade Group
CD	Canadian Forces Decoration
CER	Combat Engineer Regiment
CF	Canadian Forces
CFB	Canadian Forces Base
CFS	Canadian Forces Station
CIMIC	Civilian Military Cooperation
CN	Canadian National
CO/C.O.	Commanding Officer
Co.	Company
Col/Col.	Colonel
Cpl/Cpl.	Corporal
CRPG	Canadian Ranger Patrol Group
Dept.	Department
Det	Detachment
DC	direct current
DEW	Distant Early Warning
DND	Department of National Defence
DP	Development Period
DP	Displaced Person
DSO/D.S.O.	Distinguished Service Order
Ex	Exercise

LIST OF ACRONYMS

FOL	Forward operating location
Gen/Gen.	General
GG	Governor General
G.O.C.	General Officer Commanding
GPMG	General Purpose Machine Gun
GPS	Global Positioning System
Group GSM	Group Sergeant Major
HBC	Hudson's Bay Company
HBQ	Hudson Bay Quest
HMCS	Her Majesty's Canadian Ship
Hon.	Honourable
HQ	headquarters
ID	identification
IED	Improvised Explosive Device
IMP	individual meal packet
JCR	Junior Canadian Ranger
JCR O	Junior Canadian Ranger Officer
JTFN	Joint Task Force North
Ky.	Kentucky
LED	Light Emitting Diode
LCdr	Lieutenant-Commander
LCol/LCol.	Lieutenant-Colonel
LFAA	Land Forces Atlantic Area
LFCA	Land Forces Central Area
LFQA	Land Forces Quebec Area
LFWA	Land Forces Western Area
LGen	Lieutenant-General
LOSV	Light Over Snow Vehicle
Lieut.-Col.	Lieutenant-Colonel
Lt.	Lieutanant
Lt.-Col.	Lieutenant-Colonel
Ltd.	Limited
Lt.-Gen.	Lieutenant-General
Lt.-Gov.	Lieutenant-Governor
LUWV	Light Utility Vehicle Wheeled
Maj/Maj.	Major
Maj.-Gen.	Major-General
MASH	Mobile Army Surgical Hospital
Master Cpl.	Master Corporal

MC/M.C.	Military Cross
MCpl	Master Corporal
MGen	Major-General
M.M.	Military Medal
MMM	Member of the Order of Military Merit
MP	Member of Parliament
MWO	Master Warrant Officer
NCO	non-commissioned officer
Nfld.	Newfoundland
No.	number
NORAD	North American Aerospace Defence Command
NRHQ	Northern Region Headquarters
N.S.	Nova Scotia
NWS	North Warning System
NWT	Northwest Territories
O.B.E.	Order of the British Empire
O.C.	officer commanding
Ont.	Ontario
Op	Operation
PC	Patrol Commander
PCMR	Pacific Coast Militia Rangers
PPCLI	Princess Patricia's Canadian Light Infantry
Pte/Pte.	Private
Que.	Quebec
RCAF	Royal Canadian Air Force
RCMP	Royal Canadian Mounted Police
RCN	Royal Canadian Navy
Rgr/Rgr.	Ranger
RIB	rigid inflatable boat
RLO	Ranger Liaison Officer
ROWPU	Reverse-osmosis water-purification unit
RV	Rendezvous
SAR	search and rescue
Sask.	Saskatchewan
Sat phone	satellite phone
Sgt/Sgt.	Sergeant
SOVOPs	Sovereignty Operations
Tuk	Tuktoyaktuk

LIST OF ACRONYMS

US/U.S.	United States
USSR	Union of Soviet Socialist Republics
VC/V.C.	Victoria Cross
VHF	Very High Frequency
VIP	very important person
WO	Warrant Officer
Y.T.	Yukon Territory

– INDEX –

INDEX

INDEX

INDEX

INDEX